SWEET WATER HUNT

CONNIE NYE

SWEET WATER HUNT

CONNIE NYE

authorHOUSE®

AuthorHouse™ LLC
1663 Liberty Drive
Bloomington, IN 47403
www.authorhouse.com
Phone: 1-800-839-8640

Published by AuthorHouse   11/09/2013

ISBN: 978-1-4918-0895-5 (sc)
ISBN: 978-1-4918-0894-8 (hc)
ISBN: 978-1-4918-0893-1 (e)

Library of Congress Control Number: 2013914851

Scat Rap reprinted with permission from Crawdads, Doodlebugs & Creasy Greens Songbook: Songs, Stories & Lore Celebrating the Natural World by Doug Elliot; book published by Native Ground Music/song published by Fracas Music (BMI); www.nativeground.com. Words based on "Scat Rap" written by Rodd Pemble, Mary Keebler and Andy Bennett, Great Smoky Mountain Institute, 1988.

Parts of the Myrick Center Quest (written by Eric Goldfischer) and www.brandywinewatershed. org reprinted with permission from Brandywine Valley Association; property of Brandywine Valley Association/Red Clay Valley Association.

The Prophecy of the Fourth Crow reprinted from the University of Pennsylvania Museum of Archaeology and Anthropology as told by Robert Red Hawk Ruth and translated by Shelley DePaul.

To all the students, campers, teachers, and environmental educators of the Brandywine Valley whose senses of wonder and humor matched my own and encouraged me to take this journey.

And to my dad, Jack Nye

"Work is fun; it's fun that's work."

If there is magic on this planet, it is contained in water.

LORAN EISELY, *The Immense Journey*, 1957

CONTENTS

Maps and Illustrations

THE BRANDYWINE WATERSHED

PROLOGUE

Dry leaves swirled on the ground as the autumn wind blew. Clouds loomed over the bare branches of a sycamore tree. Overhead, four crows flew, cawing out sounds of warning.

A cloaked figure knelt beside the creek, dipped a hand into the chilly trickle, and released a white tennis ball. Partially submerged, bumping and bobbing its way down the Brandywine, the ball would, the cloaked figure thought, flow down the once-sweet water.

PART I

THE CHRISTINA RIVER
TO THE
BRANDYWINE CONFLUENCE

Summer

July

Chapter 1

THE SWEDES LAND

The July sun beat down on the back of Wyatt's arms as a spray of Delaware River splattered his face. He leaned farther over the railing of the historic tall ship and stared down at the water.

"Wyatt, don't lean so far over," warned his mother.

"Let him lean," said the commander of the Kalmar Nyckel. "If he goes over, it'll be one less mouth to feed." The three-masted ship lifted slightly with the swell and then eased into the trough of a wake left by a powerboat zipping ahead on its way to the bay.

"Starboard side, newfangled vessel!" bellowed the commander of the Kalmar Nyckel. "Darn the swine, taking us off course with its blasted wake!" The feather plume of his felt hat quivered. He shook his fist at the boat.

"It's not like we're actually on a course, are we, Mom?" Wyatt asked. "I mean, we're just sailing around the river, right?"

"He's an actor, honey, so he has to talk like he's back in the sixteen hundreds. It's *drama*." Meg Nystrom tousled her son's blond hair. He winced and turned back toward the water.

"There she is, ladies and gentlemen, the open water of the Delaware Bay," Commander Minuet addressed the small crowd of passengers. "Our reenactment of the first landing in New Sweden can now begin." He introduced himself to the crowd gathered on the deck, his puffy sleeves fluttering in the breeze. "I am Peter Minuit, commander of the first Kalmar Nyckel expedition, and first colonial governor of New Sweden."

"I thought the first governor was Johan Printz," Wyatt said. He caught his mother eyeing him with her "take it easy, smart boy" look.

Illustration by Stephen Johnson, *Courtesy* Kalmar Nyckel Foundation

KALMAR NYCKEL SHIP

"Johan Printz, though more well-known than I, arrived on the scene several years after me. I was the first governor of New Sweden, if only for a few months," Commander Minuet said. "You see, on a subsequent trip from Sweden, bringing more settlers to our new colony, I perished in a storm."

Wyatt nodded. "No governor of Delaware, before or since, has weighed as much as Johan Printz."

Check It Out!

"Wyatt!" shushed his mother.

The commander nodded at Wyatt. "'Big Guts' is what the Lenape Indians called him," he said. "You're a smart one, aren't you?"

Wyatt blushed and turned away. He was not used to receiving compliments from adults other than his own mom and dad. "You just march to the beat of your own drum," his father had told him many times.

Ahead lay the open water of the Delaware Bay dotted with the sails of catamarans and punctuated by the buzzing of speedboats. The Delaware Bay, Wyatt knew from his fifth-grade geography class, was an estuary where the fresh water of the Delaware River mixed with the salt water of the Atlantic Ocean. He remembered that because maps were things he actually liked learning about.

"All right then," Commander Minuet said, "Let's go settle a colony. It is the year 1638, and the New Sweden Company has employed me to lead this expedition. Our mission: to start a Swedish settlement and trade for beaver, since the Europeans had decimated their beaver populations. One needs beaver pelts for those fashionable men's hats back in Europe, you know."

Wyatt could see some of the adults trying to listen as their children tugged at them, clearly bored. Was he the only kid who didn't mind spending a summer day doing this kind of stuff?

The commander continued. "Our crew sailed in with two dozen Swedes and Finns to settle New Sweden, known to you now as Wilmington, Delaware." He turned and signaled to Captain van der Water.

"All hands on deck!" the captain hollered. "Prepare to come about!" Crew members materialized from the forecastle. They lined up along a thick rope connected to the boom and its ballooning white sails overhead.

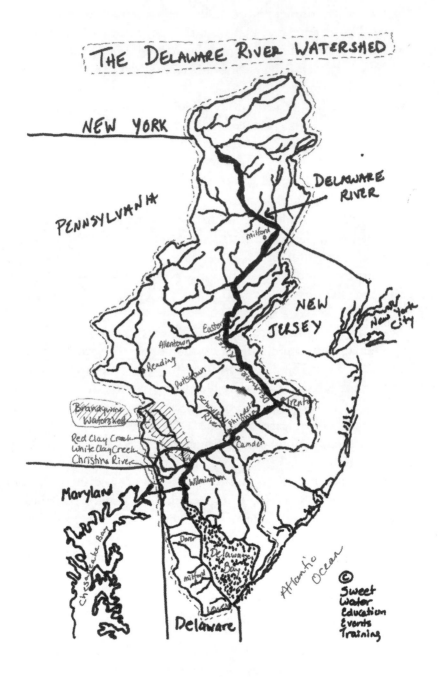

THE DELAWARE RIVER WATERSHED

"Don't just stand there, boy!" roared Commander Minuet. "Grab on and earn your keep." As the tiller man pulled back on what looked like a large joystick, the second mate bellowed, "Two, six, HEAVE!" Wyatt seized the rope alongside the crew. "Two, six, heave," he chimed in, helping to pull the sails around to ease the ship back up the Delaware River toward its confluence with the Christina River.

Map It!

The flurry of activity settled into a slower, quiet sail as the vessel glided smoothly upriver. The crew remained assembled on deck, taking in the scene to the west. As they approached the mouth of the Christina River, murmurs and whispers rose from the crew, excited and apprehensive at the same time.

"Which way are we going?"

"The river splits into two!"

"Will the natives be friendly? Will they be willing to trade?"

"The river to the west is where we're headed; it's wide and deep and the Dutch haven't yet settled there. That must be the place!"

Wyatt closed his eyes and imagined the thick, forested woodland, wild and unsettled, where the Lenape Indians trapped fox and beaver and fished for shad. When he opened his eyes again, dilapidated, industrial buildings on the banks of the Christina River and traffic on the bridge spread before him. *What a difference a few hundred years makes*, he thought. From the north, a sleepy creek eased into the Christina.

"Look, honey," his mother said, pointing ahead on the starboard side. "That's what we came to see, the place where our ancestors first landed." Wyatt looked along the bank where a rocky wharf ran alongside a flagstone plaza. A monument stood in the center of the plaza. On top of the monument, a solitary crow rested.

Check It Out!

"Yes, 'The Rocks,'" sighed Commander Minuit, who wandered over to Wyatt and Meg. "Our Fort Christina was built near the confluence of the Brandywine Creek and the Christina River. The first permanent settlement of this valley was Swedish, before the Dutch, and eventually, the English took over. Do you see the cabin?" Peter Minuit pointed to a log cabin just behind the monument. "There were trees aplenty, perfect for the Swedes to build the first log cabins in America for quick settlement to claim the land."

Map It!

Wyatt was surprised to hear that log cabins weren't invented in America, but were brought here by the Swedish. *What about apple pie and baseball?* he thought. *Were they invented somewhere other than America, too?*

"The Swedish wanted a foothold in the New World for fur trading, and this spot was perfect. Great hunting lands, a navigable waterway not yet claimed by the Dutch, perfect docking. Sweden's Queen Christina was twelve years old at the time, so I renamed this body of water the Christina River." Commander Minuit paused to let Wyatt and his mother absorb his flurry of facts. "Before I renamed this river, it was called the Minquas Kill, after the Minquas Indians. 'Kill' is the Dutch word for river, you know."

"Like the *Schuylkill*, I see," said Meg, turning toward her son. "Can you imagine, Wyatt? A queen at your age?" She paused, a glint in her eye. "I wonder if she did her homework without argument."

Wyatt rolled his eyes. "I bet she didn't have stupid Mrs. Jacobs for language arts," he muttered. He raised a pair of binoculars for a closer view of the monument. Images of Indians, portly officials, and a girl on horseback decorated the granite shaft, topped with a sculpted image of the Kalmar Nyckel ship. "Who's the girl on horseback?" Wyatt asked. "Is that the girl my age?"

"As a matter of fact, yes," answered Commander Minuit. "And you have more in common with her than her age. Her father actually raised her as a boy, since he wanted a male heir; in reality, she never would have ridden sidesaddle like a lady, as she is on the monument. When she appeared in Italy after abdicating the Swedish throne in 1654, riding horseback and dressing like a man, she caused quite a stir. A bit of a rabble-rouser, you might say. Yes, yes, I'd say you probably do have a lot in common with her."

Check It Out!

The crew of the Kalmar Nyckel moored the tall ship, securing its lines to the cleats along the dock. Impatient with his chatting mother, Wyatt scooted up the plank to the far side of the dock, where an elderly man sat on a bench. In a bucket beside him, a huge catfish swam in circles.

"Just caught him a little while ago," announced the man without looking up from baiting his hook.

"Are you going to eat him?" asked Wyatt.

"Nah. The river's too polluted to eat the fish. I just catch 'em for fun and to show 'em off."

Sticker It Red!

~ ~ ~ ~ ~

The thin man behind the counter wore an apron reading I SCREAM, YOU SCREAM. "What flavor will it be, young man?"

"I'll have vanilla," Wyatt replied. "In a pretzel cone. With rainbow sprinkles."

Meg thanked the man when he handed her the cones. "Here's enough for these and a cone for the man with the feather in his hat who just came in," she added, handing him cash.

Outside the ice cream shop, Wyatt and his mother sat at an umbrella table.

"Is it good?" she asked. She reached over to wipe away the smudge of ice cream left on her son's nose after he licked around the cone.

Wyatt nodded.

"So, we'll meet your dad, Aunt June, Uncle John, and Danni, at Brandywine Park for a picnic dinner. Then Dad will take you home and I'll head to class. I probably won't be home until eleven or so. You know, you don't have to wait up for me."

"Yeah, Mom, I'm sure I'll be fast asleep by then," Wyatt said sarcastically. He loved staying up late in the summer when his mother went to her graduate class at the university. His dad would go to bed early, so when his mother came home, she and Wyatt would hang out before bed. They would snack on pretzels and chocolate chips and sit on the deck listening to the crickets and watching the blips of light from fireflies or the flashes of heat lightning in the distance. Some nights it would start to storm, and they would count the number of seconds from a flash of lightning to the subsequent thunder. "Five seconds for every mile away," Meg had explained about sound traveling slower than light.

Try It!

Wyatt rarely saw his dad on summer mornings. While Wyatt slept late, his father left early for his shift managing the Wilmington Waste Water Treatment Plant. "His clock is changing," he heard his mom say once when his dad complained that he rarely sees Wyatt until evening. "It's a shame you have to leave so early," she added, though during the school year, she had to get up well before sunrise for the unmerciful start of the seven-thirty high school day.

"You gotta get up early to manage poop," Luke Nystrom explained one evening to Wyatt. "If you want to keep flushing, I gotta keep working," he laughed. Wyatt's mother called it gutter talk, but Wyatt always thought that his dad's wastewater jokes were funny. The day his father shared this slogan with the entire fifth-grade class on career day, however, Wyatt couldn't help but feel betrayed about their private joke going public.

"Can we pick up Benny on the way to the park?" Wyatt asked, picturing his black Labrador retriever spiraling into ecstasy as they headed over to Brandywine Park.

"Of course we can," she said. She stood up to leave just as the bells on the door of the ice cream shop jingled.

"Do I have you to thank for my delicious ice cream concoction?" Peter Minuit asked Wyatt and his mother.

"Oh, it was the least we could do," replied Meg. "Thanks again for a wonderful ride. It was very interesting."

"You're very welcome," Peter Minuit replied. "And you, sir, did you enjoy your afternoon excursion on the seas?"

"Yeah," Wyatt answered. "It was pretty cool."

"Well, you don't get any better endorsement than that," Meg said.

"Now, listen, son, I hope I didn't offend you by suggesting you may be a rabble-rouser like Queen Christina. But you do strike me as an inquisitive child with a lot of intelligence. That's a good combination. Make use of it." The commander bowed deeply, sweeping his feathered hat down low. "And stay safe in this new world. You know, it can be quite perilous," he added as he walked away.

"What's a rabble-rouser?" Wyatt asked his mother.

"It's someone who goes out of his way to stir up trouble."

Wyatt considered the term, thinking it didn't really sound like him at all. He never stirred up trouble. At least, not on purpose.

Field Trip!

Chapter 2

THE HUNT BEGINS

On the east bank of the Brandywine Creek, Wyatt and his cousin, Danni, stood in the shade of a sycamore tree. The creek water parted around large boulders and tumbled over smaller rocks. Wyatt waved to his parents on the west bank as his cousin turned a cartwheel on the grass. They had crossed the creek at the bridge to take Wyatt's dog to the "off the leash" side of Brandywine Park. Wyatt loved playing fetch with Benny in the creek, even if the signs did read PELIGRO/DANGER: WATER IS POLLUTED.

Map It! and Sticker It Red!

Wyatt flung a stick. Benny dashed down the bank and into the water. Paddling easily downstream, he retrieved the stick, then fought the current to return. He scrambled up the bank, dropped the stick at Wyatt's feet, and shook, spraying water all over the cousins.

"Benny," Danni whined, wiping droplets from her springy hair.

Wyatt threw the stick farther downstream and Benny took off. The stick floated swiftly. Benny pursued, paddling furiously, and snagged it. Wyatt watched Benny struggle to turn upstream when the rapids suddenly swept him out of sight.

"He's gone," Danni screamed. She and Wyatt raced along the stream bank to the point where they last saw Benny. They clamored down the steep bank, but saw no sign of the dog. Wyatt started climbing back up. "C'mon," he pleaded. But Danni stood still, on the verge of tears. Wyatt reached his hand out and grasped her dark forearm. The contrast in skin color made him think of the time during Black History Month. His teacher didn't believe him when he told the class he had an African American cousin. When his dad found out, he marched down to school and gave old Mrs. Johnson a lesson about Black History. Wyatt pulled Danni up the bank, where Benny sat calmly.

"Oh, Benny boy, you had us scared," Wyatt said. He threw his arms around the dog, barely noticing the water soaking into his T-shirt. Danni knelt and hugged Benny from the other side.

It was then that Benny dropped a tennis ball from his mouth.

Wyatt picked up the ball, once white, but now mostly gray and stained green in spots. Benny barked and wagged his tail. "No, no, Benny, I'm not throwing it, boy," Wyatt said. "Didn't you scare us enough already?"

Danni looked at the ball. "Yuck, that's nasty!"

Wyatt turned the ball over and saw an "X" sliced into its wool covering. A smelly liquid trickled out. "Take a whiff," he laughed and threw the ball at Danni.

She jumped aside. "I'm not touchin' that thing," she said.

Benny lunged and snatched it up. He ran a circle around the cousins and dropped the ball back at Wyatt's feet.

"OK, Benny, just not in the creek this time." Wyatt pitched the ball into the field, away from the water.

Benny chased and pranced back with the ball in his mouth. He dropped it at Wyatt's feet. When Wyatt picked it up again, he noticed that it held an odd shape, not exactly round. Something was lodged inside. He split the ball further and peered inside.

"What is it?" Danni asked.

"Something's stuck inside." He reached in and pulled a slimy block of wood out. "It's been carved," he said, rotating the wood and scraping some of the algae away. "There are words carved into the wood."

Danni drew near. "What does it say?"

Wyatt squinted at the markings. "'Sad,'" he read as he turned the cube. "And 'bad.'"

"I think we should go back, Wyatt," Danni said. "I'm getting kind of creeped out."

"Yeah," said Wyatt. He shoved the cube in the back pocket of his shorts. "C'mon, boy," he called to Benny, who picked up the tennis ball and sat obediently to be leashed.

~ ~ ~ ~ ~

Meg Nystrom beckoned to Wyatt and Danni as they ran across the bridge and crossed the blue-gray cobblestone. Benny trotted alongside.

"Mom, Dad, guess what we got," shouted Wyatt.

"Yeah, Benny found it," added Danni.

Luke Nystrom held up his arm like a crossing guard stopping traffic. "Whoa, whoa, whoa, kids, slow down before you knock someone over." He gestured to the left. "Like your mom's professor."

Wyatt glanced over to see the back of a pear-shaped woman in a long, fluttering skirt. She waved her arms as she talked to Danni's mom and dad.

"So, champ, what have you got?" said Luke.

Wyatt's eyes gleamed as he reached into his back pocket. "Benny found an old tennis ball and this was inside it. Look!"

The pear-shaped woman wandered over with Danni's mother. In contrast to Aunt June's black crop and clear, ebony skin, the woman's gray hair and lined face made her look old. Uncle John followed. His large frame and full head of red hair reminded Wyatt of a Viking. All Uncle John needed was a set of horns and a beer stein.

"Wyatt, Danni, this is Dr. Waters, my professor at the University." The professor smiled. "You can call me Dr. Flo, as in Florence."

"Hi," the cousins said.

Dr. Flo looked at Wyatt. "What have you got there?"

Wyatt held up the block of wood to show her. "Benny got it from the creek. It has a message," he said. "About something bad."

"And something sad," added Danni.

"Where exactly did he find it?" Dr. Flo's voice was suddenly stern. For someone he just met, this Dr. Flo sounded a little too much like one of his teachers. Wyatt hesitated, unsure whether he should tell her more.

"I'm sorry, I didn't mean to alarm you," Dr. Flo said, composing herself. "But, it's just not every day that someone pulls a message from the creek about something bad and sad."

Wyatt's mother intervened. "Dr. Flo teaches Wildlife Ecology at the university," she explained. "She spends a lot of time studying the Brandywine."

Dr. Flo pulled a notepad and pencil from her skirt pocket. "If it has a message with carved words, you might be able to read it better if you do a rubbing. Try washing off your block of wood really well,

16

and letting it dry thoroughly. Then rub it. Like this." She walked over to a sycamore and pressed a small paper against an area of bone-white bark, where an outer layer had peeled off. "Some like to call this the camouflage tree," she explained as she rubbed the graphite against the paper. An image resembling the letter "Y" materialized. On the bark, the letter was almost imperceptible. "See what you can decipher from your woodblock," she said.

Try It!

Down by the creek, Wyatt and Danni squatted side by side, taking turns scrubbing the woodblock clean of algae. "I'm glad your mom said you can sleep over," said Wyatt. "We can do the rubbing as soon as it's dry, which should be later tonight."

The cousins linked pinkies in agreement.

During the ride home, lightning flashed across the night sky, illuminating Riddle Avenue. "One, hippopotamus, two hippopotamus, three hippopotamus, four hippopotamus, five hippopotamus, six hippopotamus," Wyatt counted. Thunder rumbled in the distance, more than a mile away.

~ ~ ~ ~ ~

"Let's go, Danni," Wyatt whispered on the other side of the bathroom door. Danni emerged, wearing Wyatt's red shorts that fell to her calves, and his extra-large monster-truck T-shirt. The outfit didn't help her reputation as the smallest kid ever about to enter the fifth grade. "You look like a clown!" he laughed.

"It's not my fault. If you had asked me to sleep over before we went to the park, I'd have my own clothes," she replied, pretending to be offended. She glanced at the brass doorknob and broke into laughter. The reflection in the shiny knob distorted her image, making her look both super short and super fat.

Try It!

"Goodnight, Dad" Wyatt called downstairs.
"Goodnight, Uncle Luke," Danni said.

"Going to bed before me?" Wyatt's dad called up. "That's unusual."

"Mom said not to wait up for her. Besides, we're really tired." Wyatt looked at Danni mischievously.

In his bedroom, Wyatt pulled the block of wood out from under his sleeping bag. Danni retrieved several sharpened pencils and some blank paper from under hers. The cousins sat at Wyatt's desk, which overlooked the Brandywine Creek, and took turns rubbing the carved surfaces of the wood. They worked silently, keeping their promise not to read the words aloud until all of the sides had been rubbed.

Rain began to splatter against the window. Lightning flashed. The bedroom light flickered. "One, hippopotamus, two hippopotamus . . ." Thunder cracked for several seconds.

"Quick, close the window. That was only two hippopotamuses," Wyatt yelled, pulling the desk back so Danni could reach it. "It's getting close!"

Raindrops pelted the window. Danni pushed it closed with a thud. She turned back to Wyatt and picked up the papers with the rubbings. "Are you ready?"

Wyatt nodded. They sat on the sleeping bags and held the papers between them.

Danni read first. "It's just the letter '*Y*.' And '*ch44*.' What could that mean?"

Wyatt shrugged and read the next paper. "This one says '*Gladness,*' but there's a slash through it."

Danni read next. "'*Sadness.*'"

"'*Badness,*'" Wyatt continued.

Wyatt closed his eyes, pausing before looking at the last paper in Danni's hands.

"'*Who will help to stop the Madness?*'" she read aloud.

Lightning cracked, thunder crashed. The room went black.

Field Trip!

Chapter 3

THE RIDDLE IS REVEALED

"Missing something?" Meg asked.

Wyatt and Danni were groggy. They were up unusually early for a summer morning.

Dr. Flo stood by the kitchen sink. Wyatt glanced at the table where the sheets of pencil-rubbed paper lay.

"I asked Dr. Waters to stop over because of your discovery last night. I found those papers and was concerned," said Meg.

Wyatt wondered when his mother had come into his room and taken the papers. He had promised himself he'd hold on to them through the night. The last thing he remembered was Benny stretched out at the end of the sleeping bags, like a guard, and Danni asleep, her back pressed against his, the way she slept sometimes after watching a scary movie.

Dr. Flo smiled. "It's some mystery you've stumbled upon, I must say. Very exciting. And serendipitous that, to get here, I walk along *Riddle Avenue*."

"Riddle Avenue down the street?" Danni asked, reaching into the cabinet for bowls. Wyatt brought milk and a box of cereal to the table. "You live that close?"

"Close enough. I was on my way to the park for my morning hike, so I thought I'd stop by. So, you've made some progress on your woodblock."

"Yeah, but now what do we do?" asked Wyatt. He took a spoonful of cereal, dripping milk onto the table.

"Why don't we discuss that after your breakfast? Benny looks like he could use a walk." At the sound of the word, Benny froze and cocked his head.

"Dr. Flo, you have to spell the 'w' word, not say it. Now we *have* to go," said Wyatt. "Look at him." Benny circled the table, eager for a walk.

"Why don't we take him over to Alapocas Woods? Would that be all right if I take the kids to the park, Meg?"

Map It!

"Of course," replied Meg. "And I'd join you, but I've got some reading to do for your class."

"Pity," said Dr. Flo.

"Yeah, pity," said Wyatt. "Doesn't homework stink?"

~ ~ ~ ~ ~

The hike from the elegant condominiums to the Alapocas Woods was a short one, but for Wyatt, it was like a trip through history. They passed the revitalized condominiums and offices of the old Bancroft Mills site, crossed the wooden footbridge over the Brandywine Creek and entered what seemed like another geologic era. Massive walls of grayish blue stone bordered the old quarry and Indian camp. Paved paths trailed uphill, disappearing into the lush growth of beech, hickory, and tulip trees. The children, dog, and woman hiked up a trail to a bench and sat.

"What's the name of the local baseball team?" Dr. Flo asked. "I suddenly can't remember. Darn this old age."

"The Blue Rocks," said Danni, proud to be quicker than Wyatt with an answer to a sports question.

"Did you notice anything about the rocks in this park?" Dr. Flo asked, leading them on.

"They're big," answered Wyatt.

"And people must like to climb them," added Danni as two men passed by carrying crash mats and helmets. She had seen a sign for the park's climbing area, but hadn't realized how seriously some people take their rock-climbing.

Dr. Flo pointed downhill at the boulders. "Look. What colors do you see?"

"Gray," they both answered.

"Look more closely. Really discern the colors, distinguish them."

Danni and Wyatt looked puzzled, but concentrated on the task. "Dark gray?" offered Wyatt.

Danni's eyes widened. "Blue," she said. "The rocks are kind of blue. Like the Blue Rocks."

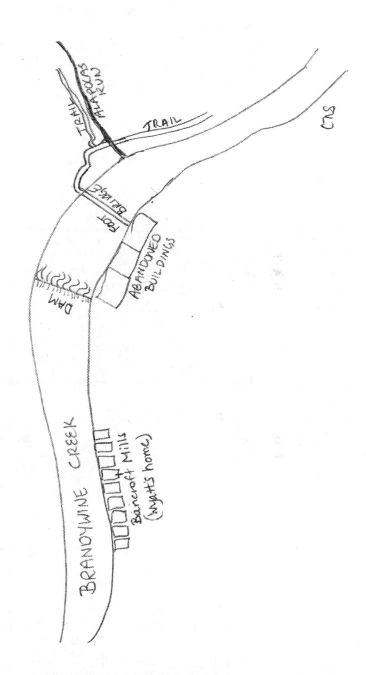

ALAPOCAS STATE PARK

Dr. Flo smiled. "Exactly. Sometimes you have to look carefully to make connections and understand things. So now you know where the baseball team got its name. It's from the rock found here in Wilmington, also known as 'Brandywine granite.' Now, let's take a look at your woodblock, shall we, and see if we can start making some connections."

Wyatt pulled the woodblock and paper rubbings from a side pocket of his cargo shorts. He laid the papers on the ground. Holding the wood in his hands, he read the words aloud, though, by now, they were committed to memory.

~~Gladness~~
Sadness
Badness
Who will help to stop the Madness?
Y
ch44

"May I see it for a minute?" Dr. Flo asked.

Wyatt handed the woodblock to her. With eyes closed, she rotated the block around, feeling all six sides carefully. "So, what do you think this woodblock could mean, and where do you think it came from?" she asked. "Now, think carefully."

"Well," said Danni. "Benny found the ball in the creek, and the woodblock was in the ball. It was slimy, so it was probably in the creek for a while."

"Good deduction, Danni," said Dr. Flo.

"And it must have come from upstream," Wyatt added. "Because things only flow downstream."

"Now you're cookin'!" Dr. Flo clapped her hands with satisfaction. "So, how do you find out more about where this came from and what it means?"

"Start checking things out upstream?" Wyatt said.

"That's a great idea, but what do you mean by 'checking things out'?"

Danni and Wyatt looked at her blankly.

"Let me show you something." Dr. Flo led the kids along a dirt path to a small stream running downhill to the Brandywine. "This is a tributary of running water called Alapocas Run. You can see that it's a small creek that flows into the larger Brandywine Creek. It's a pretty little tributary, isn't it?" The kids nodded. "Now, who do you think uses it?"

Map It!

"Deer maybe," Wyatt offered.

Dr. Flo pointed out a muddy spot along the tributary where tracks were imbedded. She placed two fingers into one of the impressions. "Two fingers clear, the foot of a deer." In the mud, she outlined the impression of a small skinny-fingered hand. "'Raccoon," she said. "It looks like a human handprint. Now, sit still, close your eyes, and listen."

Check It Out!

The rapid-fire trill of the cicadas were the first and most obvious, but soon, other sounds less frequent but more pleasant joined the medley: the laughing of a flicker, the banjo twang of a green frog, the rat-a-tat-tat of a woodpecker, the twittering of an oriole and, suddenly, the shriek of a young red-tailed hawk soaring overhead.

Check It Out!

"A lot of birds," said Wyatt.

"So many creatures are connected to this creek, either by using it directly, like deer and raccoon, or by using it indirectly. The birds are around here for a reason." Dr. Flo knelt by the creek and dislodged a rock. "Look closely."

Wyatt and Danni knelt by her side. A small creature the length of a fingernail wiggled its way across the bottom of the rock.

"A crayfish," Danni said.

"No, Danni, actually, this is an insect," Dr. Flo corrected. "This is the nymph stage of a mayfly. Think of a nymph as teenager. Soon, the mayfly will grow mature enough to have its own family. It will come out of the water, shed its exoskeleton, and fly away. So why do you think the birds are around?"

Check It Out!

"They eat the flying bugs," Wyatt guessed.

"Absolutely!"

"What's that?" Danni pointed to a whitish splotch oozing its way across the bottom of the rock.

Dr. Flo peeled the creature from the rock gently. "Ah! One of my favorites. It's a planaria, a flatworm. Look closely and you can see its two eyespots that help it sense light. Did you know that if you cut this organism in half, each will grow back into a full planaria again?"

Check It Out!

"Like a starfish," Danni added.

"Yes, like a starfish. Now, go on, see what else you can find."

Wyatt picked up a large rock. "Nothing," he said, discouraged.

"Have some patience, son," Dr. Flo coached. "Look for movement."

Wyatt peered closely at the rock. "Wait, here's something." He pointed to a tiny creature emerging from a small sanctuary of sand grains stuck to the rock.

Dr. Flo gently picked off the caddisfly and placed it in Danni's cupped hands. "Just add water," she said. She scooped up some creek water and transferred it to Danni. Immediately the caddisfly wriggled. "A case-building caddisfly is a goodie. All of these organisms tell us something about the quality of the water in this little tributary. If they are big enough to see with just your eyes, they are *macro*." She ran her thumb along her backbone. "If they don't have one of these, they are *invertebrates*. Some macroinvertebrates can only live in very clean water. Others can live in so-so water, and others can live in anything, polluted or clean."

"What kind of water do these guys like?" Danni asked.

"This kind of caddisfly and the mayfly require very clean water, by Mother Nature's standards. The planaria is far more tolerant and can live in polluted water."

"So what does it mean about this water?" Wyatt asked.

"We'd have to use a more scientific approach to be accurate about the water quality," Dr. Flo explained. "But it's not a bad sign if we've picked up two rocks close to each other and found caddisfly larvae and mayfly nymphs. Let's head downstream."

Try It!

Wyatt ran ahead and Benny trotted alongside him. "I'll beat you," he called back to Danni, who ignored him and skipped. Dr. Flo ambled down the trail until she caught up with the children. They stood still, watching the spot where Alapocas Run merged with the powerful current of the Brandywine Creek. Just upstream on the Brandywine was a dam, behind which the water moved slowly in comparison.

"A convergence of lotic waters," Dr. Flo said. The kids looked at the water, puzzled by Dr. Flo's language.

"English, please," said Wyatt.

"Oh, sorry," replied Dr. Flo. "Lotic. Think locomotion. Chugga, chugga, chugga, chugga, chugga, chugga, chugga, chugga, whoo hoo, whoo hoo."

Wyatt and Danni exchanged looks.

"Lotic; it means water that moves. Like a locomotive moves. You do know what a locomotive is, don't you?"

Both kids nodded.

"A convergence is where one stream or tributary flows into another one, so this is a convergence of lotic waters."

"OK," said Wyatt. "But does this tributary make a difference in the Brandywine? Because it's a lot smaller than the Brandywine."

"That's a great question, Wyatt," said Dr. Flo. "If you're really interested, I'll have to show you something that may help you figure that out yourself. In the meantime, if you dropped a leaf into the tributary, do you know where it would go?"

"Oh, yeah." Wyatt recalled the prior day's voyage on the ship. "It would get carried to the Brandywine, which goes into the Christina River. From there, it would go to the Delaware River."

"And then?"

"The Delaware Bay, and then the ocean."

"Which ocean would that be?"

"The Atlantic," said Danni who had been feeling left out. "Carried there by lotic waters," she added, smiling.

"Try it," suggested Dr. Flo. "I'll keep Benny with me."

Check It Out!

Danni and Wyatt made their way up the tributary. They dropped a large tulip tree leaf into the current. The leaf floated downstream for a second before it got pinned behind a rock, the flow of water pressing and holding it in place.

"Well, sometimes that happens, too," said Dr. Flo. "Things can get trapped for a *long* time."

Wyatt felt the tennis ball in his pocket. *Could that be why a white tennis ball would turn gray?* he thought.

Danni pulled the leaf out from its trap. It floated downstream to the Brandywine, where the swifter current carried it out of sight.

~ ~ ~ ~ ~

The group crossed back over the bridge. Midway, Danni stopped to look down at the creek, flanked by classy condominiums upstream and boarded-up, graffitied buildings downstream.

"Watch this," called Wyatt. He leaned over the side of the bridge and spit. A white wad hit the creek with a splat. "It's on its way to the Atlantic Ocean!"

"You're so gross," said Danni.

Dr. Flo shook her head. "Someone always lives downstream, you know."

"It's just a little spit," said Wyatt.

"Imagine if every boy who crossed a bridge spit into the Brandywine thinking that 'it's just a little spit.' What do you think would happen?" Dr. Flo had a way of explaining things without actually explaining.

"Ick!" said Danni. "There would be lots of gobbers floating down the creek!"

"Well, it does all add up, you're right about that."

"Sorry," said Wyatt.

"Apology accepted," said Dr. Flo. "Just as long as we learn something. After all, that's the purpose of mistakes: to teach us."

~ ~ ~ ~ ~

"How about some iced tea?" Meg Nystrom asked when the group entered the kitchen.

"Sounds wonderful. Thank you." Dr. Flo sat down. At her feet, Benny sprawled out on the tile floor under the table. "Do you have a map of this area?"

Meg opened a drawer stuffed with maps. She sorted through them until she found the map of Alapocas Woods and Park.

"Look, kids," said Dr. Flo. "Try to find where we are on this map and where we walked."

Wyatt traced the trail with his finger. "This must be the bridge we walked over. And here's the tributary, Alapocas Run. But I don't see where the dam is on this map."

Dr. Flo peered down through her bifocals. "That dam should be on this map. It's actually the reason you have a roof over your head."

The children looked puzzled. Dr. Flo continued. "Stop looking at the map for a minute. Now, Wyatt, tell me where you live."

"I live here," he answered. *Duh*, he thought.

"Well, I know that. I mean your address."

"Bancroft Mills, Wilmington, Delaware."

"So," Dr. Flo continued, "remember our example of the Blue Rocks? Why would your condominium be named Bancroft Mills, and what would a dam have to do with your roof?" Dr. Flo winked at Meg, who smiled in return.

"It was a bank," shouted Danni, thinking the name Bancroft might actually be Bank-roft. The others looked at her. Dr. Flo shook her head.

"A mill! This used to be a mill," exclaimed Wyatt. "Mills always had dams."

"Bingo," chuckled Dr. Flo. "Now you're starting to think like a detective. Which is what you'll have to do if you want to make progress in solving the mystery of your woodblock."

Meg smiled at Dr. Flo. "I remember reading about this area when we were thinking of buying. These buildings used to house the workers of the Bancroft Cotton Mill that functioned until rather recently, I believe."

"At one time, it was the largest cotton finishing mill in the world," said Dr. Flo. "Wilmington actually had lots of mills along the

Brandywine through America's first few hundred years, because of the water power."

A buzzing from Wyatt's pocket halted the conversation.

"You know," Dr. Flo continued, "before there were things like electricity and batteries."

Wyatt tapped rapidly on the phone's keyboard.

"And texting," added Meg.

Field Trip!

August

Chapter 4

WHAT'S A WATERSHED?

"Welcome to my kooky hoose," Dr. Flo called out to Wyatt. She rose from the swing on her front porch.

"Don't you mean *house?*" Wyatt asked. He walked up the creaky front steps. He had to agree, the old house did look kooky. The peeling paint on the porch wasn't that strange, but the sculptures of giant bugs were definitely weird. And the large plastic barrels at the end of the downspouts? That was a new one. On the other hand, beautiful bursts of flowers were everywhere: spires of red, cascading yellows, clusters of purple. Butterflies of various sizes and colors, yellow and black, orange and blue, fluttered gracefully. A strange hum whirred past and, for an instant, Wyatt saw a hummingbird hover over spikes of lavender flowers before it flew off.

"Oh, no, I mean *hoose,*" Dr. Flo answered with a mischievous grin. "As in 'Home Of Organic and Sustainable Endeavors.' I humbly invite you in."

Wyatt followed Dr. Flo through the screen door. The living room was sparsely furnished and had a musky odor, reminding Wyatt of his overnight camp cabin. Over the worn couch, the portrait of a stately man stared across the room.

"Who's that?" Wyatt asked.

"That's Uncle Clayt," said Dr. Flo, gazing at the portrait. "He wasn't my real uncle, but I thought of him that way. He was a good friend of my father's and a very interesting and fun man. When he died, back in 1986, his daughter thought I might like his portrait."

Looking at the portrait, Wyatt didn't think he looked like a fun man, but decided not to say anything. He scanned the rest of the room. Lining two of the walls were shelves filled with books and artifacts: "Deer Skull, Poconos, 1950"; "Gnawed Antler, Alapocas Woods, 1976"; "Geode, Moab, Utah, 1980"; "Snakeskin, BVA, 1997." The shelves held oddities that Wyatt had never seen anywhere but in a museum, and certainly never in someone's living room.

THE KOOKY HOOSE

"Go ahead, touch. They won't bite. Maybe." Dr. Flo left the room, leaving Wyatt to explore.

Tacked up on the fourth wall was an enormous map, pieced together like a puzzle. Wyatt stood before it. The map reached from floor to ceiling and spread just a bit wider than his outstretched arms. YOU ARE HERE: WAWASET PARK read a small, lettered flag stuck to the bottom. Wyatt searched the map to find his own home, but the squiggly brown lines and unidentified streets were confusing.

Map It!

Dr. Flo came back with a plateful of cookies. She flicked on a floor lamp directed toward the map and traced a line from her home in Wawaset Park, past Riddle Avenue, toward the Brandywine. "You live here," she said, pointing to the street Wyatt recognized as his own. She opened a drawer in the coffee table and pulled out a small sticky-backed flag. She handed it to Wyatt to press on to the map, designating his home along the Brandywine Creek.

Map It!

"This is a weird map," Wyatt commented. "There aren't any street names. How do you use it?"

Dr. Flo pointed out Bancroft Avenue and Delaware Avenue, clearly labeled. "Some streets are labeled, but not most. This is no ordinary map."

Wyatt pointed near the intersection of Bancroft and Delaware to a blackened shape labeled Highland Elementary School. "Hey, that's my old school."

"Yes, indeed," said Dr. Flo. She pointed out features on the map that Wyatt recognized: the Delaware Art Museum, Rockford Park, the Wilmington campus of the University of Delaware.

"Where's Warner Elementary? That's Danni's school."

"Right over here, by Brandywine Park." Dr. Flo tapped on the map. "And here," as she dragged her finger north on the map, "is where your mom is doing research today while you and I spend the day together." She pointed to the spot labeled Winterthur Museum and Gardens.

Map It!

"If this isn't a street map, what kind of map is it?" Wyatt asked.

"It's called a topographic map or topo map for short. It identifies lots of things, but it's unique because it indicates elevation. In other words, hills and dales. Or should I say valleys," Dr. Flo added. "Look at this." With a pen, she drew sets of circles going up each of the base knuckles of her left fist. "Think of my knuckles like a mountain range. These circles on my knuckles are drawn to connect elevations, or heights, that are the same. The smallest circle is at the highest point on each knuckle. Now watch." She flattened her hand on the table. "The circles flatten out, but they still indicate the elevations. It's the same on a topo map." She pointed to the Alapocas Woods section of the map. "Each brown line connects common elevations. A topo map is a two-dimensional representation of a three-dimensional landscape."

Try It!

Wyatt followed a brown line on the map as it squirmed its way along the Brandywine.

"From the creek," Dr. Flo continued, "cross over brown lines until you get to a closed brown circle."

Wyatt did as he was told. "Some lines are really close and some are far apart."

"Good observation," said Dr. Flo. "Do you remember hiking there last month? After we crossed the creek, we hiked on a path along Alapocas Run. Did we go uphill or down?"

"Up, definitely."

"Yes. The closer the lines are on the map, the steeper the hill."

"So the farther apart they are, the flatter it is?"

"Yes. So, reading a topo map enables you to understand elevation, which is essential to understanding the 'just a little spit' theory. Do you remember that on the bridge?"

"I said I was sorry," said Wyatt.

"Yes, yes, but now comes the real learning part from that incident." Dr. Flo changed tactics. "Why didn't you want to go with your mother to Winterthur Gardens?"

"Because it's boring."

"Oh, no, no, no! There are the Enchanted Woods with the fairies and trolls. And the gardens and meadows and forests are lovely. And they are so important, not just for beauty, but for the environment." Dr. Flo took a blue marker out of the drawer. "Now, find Winterthur again. This time, trace all of the blue lines around Winterthur with this marker."

Try It!

Wyatt did as he was told. All of the blue lines connected to something called "Wilson Run" which connected to the Brandywine Creek. "Are all the blue lines streams?" he asked.

"Yes, and what is important is which way those streams flow." Dr. Flo paused. "Think of your tennis ball. If someone dropped it into one of the streams, which way would it go? To figure this out, you have to find nearby high points first, then think about which way water travels." She took the marker from Wyatt and identified several high points close to the blue stream lines. She placed an "X" on each of the closed brown circles. She encouraged him to continue until he had created a dot-to-dot shape around Wilson Run and its tributaries.

Try It!

"If the blue lines represent water, what does the area *around* the water represent?" she asked.

"Land?"

"Of course. That's where high points are, on the land. This area bounded by the 'Xs' designates the area of land that drains into Wilson Run, so if it rains . . ." Dr. Flo paused.

"The rain water goes into Wilson Run?"

"Exactly. Any water that doesn't absorb into the ground runs off. One way or the other, the water that lands inside the area that we outlined will end up in Wilson Run. This is known as a watershed, and it's important because the *use of the land* affects the *quality of the water*."

"Isn't a shed a building?"

"A building shed stores things, so you can think of a watershed that way if you think of the land as a way to store water. But think of your dog, too."

Wyatt thought about Benny at home alone today. Clumps of his fur were always all over the house. He didn't really care so much, but his mother sure did. He'd have to vacuum when his mother could no longer tolerate the mess. "Like shedding fur?" he asked.

"Think of the shedding fur as rainwater, Benny's body as land, and the floor as a source of water, like a lake or an ocean or Wilson Run. Do you see how the rain would be 'shed' over the land and end up in the Wilson Run?"

Wyatt smiled. "Yeah, I get it."

"We call any area of land that sheds or drains into a specific body of water a watershed. So what should we call the watershed around Wilson Run?"

"The Benny Watershed," Wyatt joked. "Just kidding. How about the Wilson Run Watershed."

"Perfect. Now," Dr. Flo said, the lesson obviously not over. Wyatt didn't mind though. Dr. Flo had a way of teaching him that felt more like figuring out a riddle than sitting through school. "See the brown number on the brown line near the Brandywine?"

"Two hundred," Wyatt said.

"Yes. The elevation or height there is two hundred feet. Now, look to see a brown number on a brown line close to the far end of Wilson Run."

"Three hundred seventy."

"Good. Now if water travels . . ."

"Downhill," Wyatt answered enthusiastically, sure of himself.

Dr. Flo nodded. "Then which way is Wilson Run going, or, to put it in terms you might better relate to, which way would your tennis ball flow?"

"It would flow to the Brandywine because Wilson Run is higher than the Brandywine."

Dr. Flo smiled. "A regular Einstein, you are. And then where would it go?"

Wyatt traced Wilson Run to the Brandywine and then downstream toward Wilmington. "Hey, it would go right by my house." He continued to trace the stream. "And then to Brandywine Park where Benny found it."

Try It!

"Hm," Dr. Flo commented. "So the ball must have come from somewhere upstream." She gestured to the entire wall map.

Wyatt grimaced. "There sure are a lot of blue lines upstream."

Dr. Flo nodded. "Well, for now, let's just stick with seeing where the water goes, which is where your tennis ball would have gone if Benny hadn't found it."

Wyatt continued to trace the Brandywine to the spot where it merged with the Christina River. On the map he read "Fort C Park."

"Is that where the statue of the ship is?" he asked.

"Exactly. Close to where the Brandywine Creek and the Christina River come together at their confluence. How did you know that?"

"Me and Mom," Wyatt started.

Dr. Flo interrupted. "Mom and I," she corrected.

"Mom and I went on the Kalmar Nyckel last month, and they showed us that from the ship. It's where the Swedes first landed in the area."

"Did they talk about why they landed at that particular spot?"

"Yeah, it was because those Dutch guys didn't settle there yet and they wanted a place to get beaver to make hats." Wyatt surprised himself by how much he remembered from Commander Minuit.

"Yes, and it was a great spot, not yet discovered by other Europeans, where they could hunt for beaver and load the furs right onto the ships, which could then sail straight to Europe. You know," Dr. Flo continued, "I wonder how they survived all those hardships back then without these." She turned to the cookie plate. "I think we are deserving."

"I thought you'd never get to that," Wyatt laughed and bit into the large cookie with its chunks of dark chocolate and pecans. "Pretty good, for an oatmeal cookie," he said.

Try It!

"The secret is a surplus of chocolate. It's what keeps me going strong at seventy-five."

The two sat quietly enjoying their cookies and looking at the entire map.

"So, there are a lot of places that tennis ball could have come from," said Wyatt.

"The Brandywine is actually a small watershed. It looks large compared to the Wilson Run Watershed, but it's small compared to, let's say, all the land that drains into the Atlantic Ocean.

"That would be a lot of land, wouldn't it," said Wyatt.

"Ah, yes. Think about it. The Atlantic Ocean Watershed. Can you guess what divides America into its two major watersheds? I'll give you a hint. The two major watersheds are the Atlantic Ocean and the Pacific Ocean watersheds."

"It would have to be high points, right?" Wyatt answered.

"Yes, and what do you suppose we call those high points, the ones that *divide our continent?*"

Wyatt smiled. "I guess that would be the Continental Divide."

Check It Out!

Dr. Flo nodded, looking pleased. "Look at the Brandywine. All the little creeks and runs like Wilson Run have their own little watersheds, but they join together as part of the Brandywine Watershed. Now think, what if each blue line in the Brandywine Watershed had a bridge over it?"

"Yeah, OK."

"And each bridge had a boy on it?"

"I still don't get it."

"And each boy spit into the water?"

Wyatt thought for a minute. "All that spit would add up and go right by my house. That's *gross.*"

"You, my boy, have just identified not only parts of a watershed, but also some of the reasons for water-quality problems. Little bits of pollution add up."

"OK, OK, how sorry can I be?" Wyatt said.

~ ~ ~ ~ ~

After Wyatt watched an episode of *SpongeBob SquarePants*, Dr. Flo entered the den. "Time for a break from your 'pineapple under the sea,'" she ordered. "I'm giving you a tour."

From the den, through the living room, to the kitchen, Dr. Flo marched Wyatt around, explaining just what made her "hoose" so kooky: the twisted fluorescent bulbs; the old, black wood stove; the multicolored ceiling fan; the bike and helmet hanging against the kitchen wall. She even had a recycling *center* with stackable bins for Styrofoam, paper, aluminum, plastic, cardboard and all sorts of other junk. Then, Dr. Flo directed Wyatt to the bathroom.

"Well," she said as Wyatt exited. "Did you wash your hands?"

"Um, no."

The bathroom sink looked more like an exhibit than an appliance, and Wyatt didn't think he should use it. Painted on the bowl of the sink was a mural of a pond, complete with duckweed, insects, fish, and cattails. Over the sink, a sign read, THE LENTIC POND: POND-ER YOUR WATER FROM SOURCE TO SOURCE. A beanbag frog and a real, though lifeless, dragonfly sat in the soap dish. Painted on the wall around the sink, a water snake, raccoon, red fox, turtle, and bald eagle eyed the creatures in the pond. In the mirror, Wyatt saw only his eyes reflected. The rest of his face was absorbed into the body of a frog etched into the glass.

"I didn't really want to wash my hands with toilet water," he said, nodding toward the top of the toilet tank that had been converted into a sink. When he flushed the toilet, water from this "toilet sink" came out. Clearly, he was supposed to use this water to wash up. Then the water went down a drain into the toilet bowl to flush it. A sign over the tank read, USE RESOURCES WISELY. ONE FLUSH DOES DOUBLE DUTY (NOT DOODY). Wyatt couldn't wait to pass this one on to his dad. "And I didn't want to use the sink, since it says it's a lentic pond, whatever that is."

Check It Out!

"As for the lentic pond, you shouldn't use it to wash. Remember the lotic waters?"

"That's moving water, like a locomotive, right?" said Wyatt. "Like a stream."

TOILET SINK

"Yes, as opposed to lentic water that is non-moving or still water that you shouldn't wash with. Have you ever studied French?" Wyatt shook his head no. "Well, *lentement* means slow; lentic means slow-moving, or still water like you find in a pond. The water coming through the sink over the toilet is clean, moving or lotic, water," Dr. Flo added. "The water runs from the house pipes to the toilet-tank sink *before* it goes into the toilet to flush. That way, you can use the water to wash before it's used to flush, so it isn't wasted. Isn't it marvelous? It's my latest addition to decrease water usage. If you were a frog, you'd appreciate me not wasting water. Or if you were a snake that eats the frog, or an eagle that eats the snake."

Dr. Flo led Wyatt out to the front porch and explained the presence of the insect sculptures poking out from the shrubs. "The praying mantis, the ladybug, and the caddisfly were made by some of my former entomology students as final projects," she explained. "They had to be anatomically correct."

"What's entomology?"

"It's the study of insects. A fascinating field, I might add. And complicated. Did you know that there are over a *million* different kinds of insects? And some of them, like caddisflies, look completely different as adults from the way they look as larvae."

"Didn't we find some caddisflies in Alapocas Run?" Wyatt asked, eyeing the sculpture. "They didn't look like that."

"Yes, we did," Dr. Flo answered. "This sculpture is of the adult caddisfly; what we found in the creek were caddisfly larvae. Like a caterpillar changes into a butterfly, the caddisfly is one of many insects that transforms its shape as it grows up."

"Metamorphosis, right?" said Wyatt.

"Indeed," Dr. Flo said, smiling.

Check It Out!

She guided Wyatt to the back of the house, pointing out her quirky *endeavors,* as she referred to them: the porous driveway to keep rainwater from running to the street; the rain barrels to catch rain from the roof and use to water her gardens; the sunken rain garden planted to keep rain water from running to the storm drain; the strange-looking black solar panels angled on the roof providing hot water to the *hoose*; a large crate filled with soil, peat moss, leaves, and, hidden within, hundreds of redworms eating her kitchen food waste; and, nailed to a pole, a shallow black box for housing bats.

"You really like gross things," Wyatt commented. "Why would you want bats in your yard?"

"To eat the mosquitoes, of course! Did you know that a bat can eat six hundred mosquitoes an hour? If I had twenty bats in that bat house, how many mosquitoes could they eat in one summer night?" Dr. Flo asked.

"Six hundred times twenty equals twelve thousand," answered Wyatt. "That's a lot!"

"That's a lot? That's only one hour! If it's dark for eight hours a night, how many would they eat?"

"That's easy," replied Wyatt. "A ton!"

Try It!

Dr. Flo led Wyatt over to a small vegetable garden. She plucked a juicy blackberry from a mass of vines and popped it into her mouth. "Delicious in the warm sun. Try one."

Wyatt plucked his own, squeezing purple stain onto his fingers before popping the berry in. Warm and bumpy, its sweetness oozed from his tongue all around his mouth. He followed Dr. Flo, hunting for and tasting treats from the late summer garden: yellow wax beans, "super sweet" cherry tomatoes, and a carrot pulled straight from the soil.

"Always eat your vegetables," Dr. Flo teased. "But it's important to make sure you've had your chocolate first." She winked at Wyatt.

~ ~ ~ ~ ~

In the kitchen, Dr. Flo lined up a variety of ingredients on the table and placed them behind a glass pitcher filled with water.

"What are we making?" Wyatt asked, curious about what kind of snack could be made with powdered drink mix, vegetable oil, chocolate syrup, and salt.

"A mess, I hope," Dr. Flo replied. She retrieved a large spoon from a drawer and pointed to the pitcher. "Let's say that this is the water of the Brandywine Creek, and we are here at my house. Now that you understand how water flows from high points to a surface water source like a stream or pond, and that land can affect the water it surrounds, we're going to do a little demonstration. Think of my neighborhood. What kinds of things might go from the land around my neighborhood into the Brandywine?"

Wyatt looked at the ingredients. "Chocolate syrup?" he said.

Dr. Flo shook her head. "No, no," she said. "Think. Don't be so literal. These ingredients represent other things. Let's try this another way, without looking at the table. Imagine that I can't grow anything in my yard, so I need to do something to get my grass and flowers to grow. What might I put on my yard?"

"Oh, fertilizer," said Wyatt.

"Yes, so this powdered drink mix will do just fine as *fertilizer*. Here, pour it in."

Wyatt tore off an end of the packet and poured in the mix. The powder swirled before dissolving. It turned the water a pale yellow.

"Now, what else might go in if I washed my car in the driveway?"

"Soap," Wyatt answered and squirted some liquid detergent into the pitcher. He mixed the solution, churning up bubbles.

"And this is for chocolate milk, right?" Wyatt joked, knowing that Dr. Flo probably had something more interesting in mind for the chocolate syrup.

Dr. Flo mixed some of the syrup into a small cup of water. "This, my dear, represents the old-fashioned kind of pollution that came from homes and factories before there were adequate wastewater treatment plants and the regulations that we now have. Unfortunately, even now, factories are sometimes out of compliance with those regulations. Or sometimes the wastewater treatment plants are flooded with too much wastewater, so some of it is not treated completely before being discharged back into the creek." She had Wyatt pour some of the brown water into the pitcher. Plumes of chocolate water expanded until the entire container turned a cloudy brown.

Wyatt thought back to some of the severe summer storms when his father would come home late in the evening, exhausted from the extra hours spent dealing with too much rainwater infiltrating the aged pipes throughout the city, straining the wastewater treatment plant.

"So, that's poop water?" Wyatt asked, curious about the consequences of an overburdened wastewater treatment plant.

"Partially treated poop water. Yes, you can call it that," Dr. Flo said. "Rather disgusting, don't you think?"

Dr. Flo continued to grill Wyatt with questions until he made the proper connection between each ingredient left on the table and an actual material that could easily end up in any river or creek: salt spread on the roads after an ice storm; oil that might be illegally dumped as motor oil into a storm-water drain; trash and litter; red powdered drink mix for pesticides; soil originating from bulldozers clearing land for development, washing off from plowed fields or eroding from stream banks. Wyatt poured each ingredient into the mixture and stirred, creating a tornado of nasty brown, bubbly wastewater.

Try It!

"And now, for the grand finale." Dr. Flo reached for the chocolate syrup bottle, raised it over the pitcher, and gave it a squirt. A string of syrup hit the water, turning into a long, gooey worm on impact. "Think cows and horses and dogs," she said to Wyatt. "The creek is their bathroom."

"Cow pies?" Wyatt asked.

"Cow pies, yes, or manure, to be a little more technical. Even if the animals defecate on the soil, it can get washed into the creek in times of heavy rain. And dog waste is a huge contributor to water pollution."

"This is so cool!" Wyatt said.

"So cool? What if I tell you that this," she nodded toward the pitcher, "is what the Brandywine looked like in the 1940s. It was *madness*," Dr. Flo said provocatively. "And it can happen again to our Brandywine if we don't manage those pollutants. Forests, gardens, and open spaces like Winterthur, where your mother is today, help to filter out those pollutants before they enter our streams. And my rain garden, rain barrel, and plantings all help to keep runoff from my roof and property from entering the storm drain and running to the stream."

"Oh." Wyatt's enthusiasm wavered. "Maybe it's not so cool. All that pollution is actually kind of gross."

"Gross indeed. I was about your age when the Brandywine looked something like this." Dr. Flo pointed to the pitcher. "And it wasn't just gross. It was dangerous for those little critters, those macroinvertebrates like the ones we found in and around Alapocas Run, and all of the life that depends on them."

Wyatt thought about the connections between the pitcher, now full of polluted water, and the caddisflies, mayflies, and planaria he and Danni had found in the stream in Alapocas Woods. He thought of the mammals, birds, reptiles, fish, and amphibians that relied on the macroinvertebrates. *Could pollution have something to do with the madness carved into the woodblock?*

"Everything is connected," Dr. Flo continued, as though reading Wyatt's mind. "Now, let's get cleaned up. Your mother should be here soon to take you home."

Following Dr. Flo's instructions, Wyatt poured the concoction from the pitcher outside, into the rain garden. "It won't harm the earth," she told him. "The garden will actually filter out most of those *pollutants*."

He watched the dirty water seep into the soil and disappear. *Madness,* Dr. Flo had said of the polluted water. *Was that just coincidence? Or does she know something about where that tennis ball came from?*

Field Trip!

Chapter 5

BACK TO SCHOOL

Wyatt felt restless. Lying in bed, he worried about missing the bus on the first day of middle school. *What will my teachers be like?* he thought. *There'll be so much homework. Who will be on my sixth-grade team?* Though his new middle school was less than a mile from Highland Elementary, it felt like he was about to enter a different world.

"I can't sleep," he complained to his mother after knocking on her door and waking her.

Meg Nystrom rolled onto her back. "Neither can I, now," she said groggily. She sat up, making space, and patted the bed for Wyatt to sit. "What's the matter, honey?" she whispered, trying not to disturb Wyatt's dad.

"I don't want to go to a new school," Wyatt murmured, snuggling against her. "What if they don't *get* me?"

Meg wrapped her arms around him, grateful for the rare opportunity to comfort her adolescent son. "You'll do fine, Wyatt," she assured him. "You won't be the first kid they ever had who fidgets in class. You'll have your old friends there, and soon you'll be making new ones. You'll have new teachers who'll open up a world of new interests. There will be clubs to join and dances and all sorts of new opportunities. Think of it as the 'next great adventure of Wyatt Nystrom.' I think you'll love middle school." She looked down at her son lying against her, asleep.

Fall

September

Chapter 6

FALL FIELD TRIP

Wyatt followed his classmates as they spilled out of the white stone building of the Alexis I. du Pont Middle School. The students of Team A clamored onto the school bus. Wyatt made his way to the back and sat down beside his newest buddy, Rob.

"I heard this field trip is so boring," said Rob.

"At least it's a day off from school," Wyatt replied. Though school had only been in session for three weeks, he was looking forward to a break from the rapid pace of middle school: the crowded hallways, the rowdy lunchroom, the fickle locker combination that seemed to work only when he wasn't in a hurry, which was rarely, since there were only three minutes between classes. The adjustment to middle school wasn't easy, but Wyatt seemed to handle it about as well as every other sixth-grader.

The bus left the parking lot and grumbled its way through a right-hand turn. Wyatt knew the ride would be short, since the Hagley Museum was only a mile away.

Map It!

But the ride was long enough to get zapped by Rob, who grabbed his hand and scribbled the name of a sixth-grade girl on his palm. "Three minutes," he wrote on top of his hand.

"Three minutes?" Wyatt said. "I can't look for three minutes?" Wyatt knew the rules of the game, but had never been zapped before.

"Any less, and you have to ask her out," said Rob.

Wyatt sighed. He grinned and pointed out the window. "Look; it's the Great Pumpkin," he shouted, trying to distract Rob so he could sneak a peek at his palm.

Several kids turned to look out the window at a Great Pumpkin-less landscape, but Rob turned back in time to catch Wyatt glancing at his palm. "Now you *gotta* ask her out!" He laughed as the kids at the back end of the bus quieted.

Wyatt hesitated. "Jennie," he called softly. No response. Rob gave Wyatt an urgent look. "Jennie!" Wyatt called out to the middle of the bus. His pale complexion reddened as Jennie turned to face him. Dozens of sixth-grade faces turned to watch. "Will you go out with me?"

"No way," Jennie answered. Her light-brown ponytail, streaked with purple, whipped around as she huddled back down with her girlfriends.

"Ouch," said one of the boys seated in front of Wyatt and Rob.

"Sorry, man," Rob said, punching Wyatt's shoulder lightly.

Wyatt turned away. "Who cares, anyway?" he said. He looked out the bus window. Two crows rested on the entrance gate to the Hagley Museum. Standing nearby, a disheveled man stood, looking lost.

"Did you see that guy?" Wyatt asked Rob.

"Who, the Great Pumpkin?" Rob said, not even bothering to look.

~ ~ ~ ~ ~

"Please remove all metal buttons, belts, keys, and zippers from your clothing," the silver-bearded man said. He wore a nametag that read DOCENT DON. The kids in Wyatt's group looked bewildered as they checked their pants zippers. "And if you have matches on you, you're fired," he bellowed. "Just kidding, kids," he continued as he watched the group of boys mouth words like, "What?" and "Huh?" "But back in the days when this gunpowder mill was operating, such infractions were grounds for dismissal. You know why? Because you might blow us all sky high! Believe me, it happened. In the one hundred twenty years that this gunpowder mill operated, over two hundred men were killed in explosions. It was *madness*." Docent Don paused long enough to compose himself. "So with that introduction," he continued calmly, "let's explore the Eleuthere Irenee du Pont gunpowder mill, now known as the Hagley Museum."

He led the group into the museum's first exhibit room. Wyatt and Rob exchanged looks as if acknowledging that maybe this wouldn't be such a boring field trip after all. "By the way, the name's Don, and I'll be your guide today as we take a little trip through history." He led the group through a set of double doors to a diorama of the lower Brandywine Valley, where long stretches of green pastureland connected hills dotted with miniature farm houses, miniature trees, and miniature people. From the highland area of Honey Brook to the lowlands of Wilmington, a waterway cut through the center of the diorama.

"How many of you have ever canoed on the Brandywine before?" Don asked, pointing to the waterway. A few hands shot up. "OK. How many of you have ever paddled upstream?" One of the boys raised his hand tentatively and quickly lowered it when he saw that no one else had raised theirs. "Why is that?" Don asked.

"It's too hard," one of the boys shouted.

"Exactly, my friends," Don said. "It's too hard, because water is powerful. You want to be carried by the water's current, not fight it." He paused. "Look at the creek; which way does the water flow?"

Wyatt began to wonder if this man was somehow related to Dr. Flo. He sounded so much like her. "Downhill," Wyatt answered, recalling his lesson at Dr. Flo's kooky hoose.

"Excellent. The reason we are standing in this spot at this time is because water flows downhill. And where water flows downhill steeply is where water is most powerful. And where water is most powerful is where water provides energy that costs *nada*, nothing, not a cent. Can you imagine filling up your car's gas tank for free? Of course not. But in this five-mile stretch of the Brandywine," he said, pointing to the lower section of the diorama, "from Chadds Ford to Wilmington, the elevation of the creek drops over one hundred feet. So, in the year 1802, if you wanted free energy to power your machinery to make, let's say, gunpowder, do you think this would have been a good place to set up shop?"

The boys nodded, animated by the mention of gunpowder.

"Absolutely, boys. Now, let's find out how."

Docent Don proceeded through a room lined with exhibits. "Take a few minutes to explore," he said. Wyatt lingered in front of the display of the Lenape Indian encampment, but left when his friends moved on to check out the models of water wheels and turbines and an actual shotgun that made use of the explosives produced long ago, during the War of 1812. Don then led them outside to the old gunpowder mill. A sluggish body of water, bounded by rock walls, reflected the reds and yellows of the leaves on the surrounding autumn trees.

"Is that the Brandywine?" one boy asked. "It doesn't look very powerful."

"That is not the Brandywine, young man," Don answered good-naturedly. "That is a manmade canal called a millrace. It takes water from the Brandywine and directs it to the water wheels of the mill. We are about to see how it all works."

The group walked along the slow-moving water, tracing its origins very slightly uphill and parallel to the more recognizable and larger Brandywine Creek. A dam in the Brandywine directed water toward the millrace. "Think of this millrace as the *Son of the Brandywine*. The dam forces some of the Brandywine Creek into the race." The group crossed a bridge and gathered around a wooden platform positioned between the race and a set of majestic stone buildings facing the creek. "Here we have," Don said dramatically, "the key to the power." He chose two of the boys to raise the gate that held back the millrace water. The rest of the boys watched as water gushed below them, dropping from the higher elevation of the millrace, hitting and turning the water wheel, and falling back down to the Brandywine.

"How about that power, boys?" The group walked over to a stone building. Two enormous cast-iron wheels were connected to the water wheel by a series of gears and shafts. They rotated heavily, just as they had two hundred years ago, to grind together charcoal, sulphur, and saltpeter. "All to produce this." Don held up a small container of black gunpowder used in the guns, cannons, and blasting explosives of the 1800s. "So that they could do this." He placed the gunpowder on a metal contraption inside a large Plexiglas box. The boys' eyes widened as Don lit a match and touched it to the fuse. A blast exploded from the gadget and the boys, caught off guard, laughed from both surprise and fear. "Now, stay in line, boys. There are consequences if you don't," Don joked.

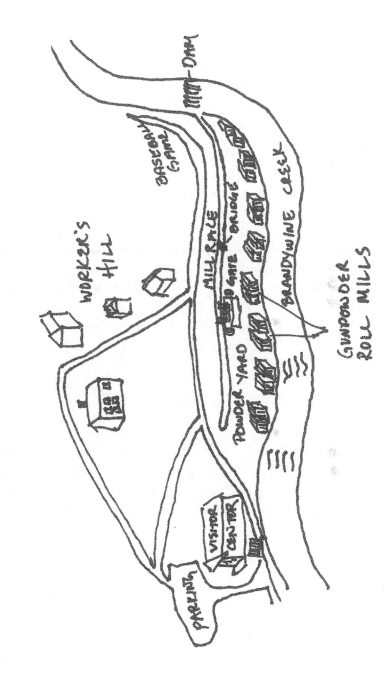

HAGLEY MUSEUM & LIBRARY AREA

Check It Out!

During the lunch break, some of the sixth-grade boys shoved each other playfully, glancing at the girls to see if they were watching. Wyatt grabbed his bag lunch from the large cardboard box and headed to the picnic table where Rob and some other boys swapped cookies and sports drinks. He looked around for Jennie, but didn't see her. Eager to join the group of boys starting up a game of stickball, he devoured his sandwich. A kid named Austin sauntered over to a designated home plate and cocked a tree branch over his shoulder like a serious ballplayer. Another boy lobbed a black walnut over the plate. As Wyatt headed to right field, Austin whacked the walnut, sending it soaring into foul territory. Wyatt chased it toward the bank of the Brandywine, where the walnut splashed into the creek and floated dam-ward.

"You stink!" one of the boys yelled at Austin, who laughed in response.

"Let's go boys," hollered Mr. Kang, the teacher who had organized the field trip. "Time to regroup!"

Wyatt watched from the bank. The dam channeled the walnut into the millrace. Snagged by a pile of debris, the walnut disappeared underwater. Wyatt turned to join the rest of the group gathering at the picnic tables, but something caught his eye. Upstream, along the bank, he saw the disheveled man he had seen earlier. His tattered, singed clothing fluttered in the breeze. The man's face, blackened with soot and dried blood, looked sad, as he seemed to search for something. He turned toward Wyatt and looked directly at him with intensity in his eyes. Wyatt shut his eyes and shook his head. When he opened his eyes, the man was gone.

~ ~ ~ ~ ~

Standing before a wall map of the Brandywine Valley, Wyatt and his group stared at the hundreds of little red triangles placed along the many branches of the creek.

"Water back then is like oil is to us today," Docent Don explained. "And milling was big business. All of those red triangles on the map were dams constructed to harness waterpower for the iron forges and the paper, textile, and gristmills throughout the valley. The Brandywine had the highest concentration of mills in the colonies. In fact, from the late sixteen hundreds to the early nineteen hundreds, over one hundred thirty mills had been constructed on the Brandywine and its tributaries." Don noted the bored looks on the boys' faces. He decided to change tactics. "Want to be a rich man?" The boys perked up. "What are you going to do to get rich?"

"Become a lawyer!" shouted one of the boys.

"No, be a movie star," offered another.

"Be a ballplayer in the World Series!" said a third boy.

"Those jobs wouldn't have made you rich a few centuries ago," said Don. "What would have made you rich? Making cloth for fine garments, like the Bancrofts did." Don pulled Wyatt aside, fished inside a canvas bag, and pulled out a tightly woven piece of tapestry. He draped it over Wyatt's shoulder. "So you would set up a textile mill. Or you might have turned rags into fine paper in your paper mill like the Gilpins." He pulled Rob from the group and handed him a piece of fancy paper, slightly crumpled from the bag. "Or, you could have been among the scores of millers like the Shipleys, the Tatnalls, and the Canbys. You could have made flour of such high quality from your grist mill that you set the world's standard with your 'Brandywine Superfine' and fetched the best prices." Don gave each of three boys a sugar cookie. The boys nodded, grinned and high-fived each other as they bit into their cookies. "All of you would have powered your mills for free, using water," Don paused. "Have you ever heard of the Shipley School? The Tatnall School? DuPont? Those are the names of families who got their start by milling on the Brandywine. And thanks to the du Ponts, not only do we have this historic mill museum, but we also have Winterthur, Longwood Gardens, and the A. I. du Pont Hospital for Children."

Wyatt had heard of most of those names and places, and had even been to some of them, but never thought about how they might be connected to the Brandywine Creek.

"My mom says that DuPont makes lots of nasty chemicals," challenged a boy named Jake.

Don nodded. "They do make lots of chemical products. But for every chemical created that may be bad for the environment, like insecticides or herbicides, there are inventions that we all take for granted, like Teflon that keeps your eggs from sticking to the frying pan, and Tyvek that acts like a windbreaker around your house. Anyone here want to be an astronaut? How about a NASCAR driver?" The boys nudged each other and nodded. "Can't do either one without DuPont inventions. Surprising, isn't it? Lots of things you take for granted can be traced back here to Hagley, where the du Pont legacy began."

Check It Out!

"That legacy should have ended before they made the du Pont Middle School," whispered Rob.

"And speaking of negative effects of industry," Don added, ignoring Rob, "many of these mills put chemicals directly into the Brandywine. Even in the eighteen hundreds, the Brandywine was fouled with sewage, cotton mill dyes, and chemicals from bleaching and processing. Not just a source of power, the Brandywine was also a place to dump wastes of all kinds that polluted everything downstream."

Wyatt thought about Dr. Flo's bathroom diagram. If the water was polluted, it would affect the macroinvertebrates as well as the fish and frogs that ate them and the eagles and snakes that ate them. *It's all connected*, he imagined Dr. Flo saying.

~ ~ ~ ~ ~

The sixth-graders spent the rest of the afternoon visiting other areas of the museum: the historic machine shop with its banging, whirring, and grinding of old tools, and the old schoolhouse where quill pens and slate boards were used long before laptops and Smart Boards. On the way out of the schoolhouse, Wyatt stopped to read a framed paper that was faded and yellowed with age.

The terrible explosion kills twelve,
Never again will we be ourselves,
The mill was destroyed,
There is left a large void,
Now the strength of our souls must be delved.

Wyatt glanced at the grainy black-and-white photograph placed below the poem. In it, twelve poker-faced workers stood stiffly. Wyatt scanned each of the faces until he reached the one on the end. A shiver went up his spine when he realized who it was. The man stared into the camera with intensity in his eyes.

~ ~ ~ ~ ~

The buses in the parking lot idled noisily. Yawning, the driver pulled on the lever and opened the bus doors with a hiss. Wyatt climbed the steps and made his way toward the back of the bus. As he passed Jennie, she looked up at him and smiled coyly. *Madness,* he thought. *Girls are total madness.*

Field Trip!

Chapter 7

CREEK PALS

Wyatt pulled out his cell phone. He figured it was Danni calling him on *her* new cell phone.

"So, when are we gonna figure out the woodblock?" Danni asked. "My teacher took us for a walk to the creek today, and did you know that there are signs all over the place that say the Brandywine is polluted? I told my teacher about what Benny found, and she thinks that the madness probably has something to do with the pollution there." The way Danni spewed out her thoughts, Wyatt figured she must have been waiting all afternoon to call him.

He had always been a little jealous of Danni when it came to school. Even though he went on the same field trips when he was in elementary school, it seemed special that Danni could simply walk to many of hers. The zoo, the park, and the Brandywine were right across the street from Warner Elementary.

Map It!

"Maybe we should go visit Dr. Flo, Danni," Wyatt said, thinking of the strange man at Hagley. "I think we need her help. Besides, you gotta see her house. It's so cool."

~ ~ ~ ~ ~

Danni stopped short when she saw the house on Blackshire Road.

Map It!

"Wow, cool," she said, approaching the caddisfly sculpture. The commotion of wilting plants, leaf litter, and bursts of orange, beige, and golden mums channeled the children toward Dr. Flo's home. Cobwebs adorned the porch and cornstalks guarded the front door. Several fake bats fluttered near the porch ceiling. From the large sycamore tree in the yard, a murder of crows cawed.

Check It Out!

Dr. Flo opened the door. "Welcome, welcome, welcome!" she sang out. "You made it!"

"Great decorations," said Danni.

"Oh, I do love Halloween. I know I start decorating too early, but I just can't help myself. That mix of nature and spookiness, I just love it!"

"What does Halloween have to do with nature?" Wyatt asked.

"Ah," Dr. Flo smiled. "I just love when we start the questions right away. What does Halloween have to do with nature? It has *everything* to do with nature. You carve pumpkins, don't you? They are one of the last crops of Mother Nature's growing season. And why do you think that witches are green?"

"Because it's scary?" said Danni.

"No, no, no," said Dr. Flo. "Scary came later. Green represents spring, the time when nature comes back to life. Isn't everyone thinking *green* these days? We think of witches as scary old hags stirring dead animal parts in a cauldron to make something poisonous, but witches are quite misunderstood. They are actually symbolic of the earth's renewal, of nature recycling dead materials back into new life. Think of a witch as Mother Nature, stirring together the dead nutrients from the end of the harvest season. The witch stirs these nutrients to ensure that, even as the days get shorter and darker, this *deadness* will come back to *life* in the spring. Witches are actually a reminder of hope and the cycling of life. Oh, I could go on and on. I sure do know a lot about witches." She winked.

"Why don't you have any ghost decorations," asked Danni.

"Oh, Danni, I don't like the idea of ghosts appearing just because it's Halloween. Ghosts are so much more than phantoms trying to scare people. I believe they are actual spirits of the dead and that they appear for a reason."

"I thought the reason was to scare people," said Wyatt.

"Oh, no!" said Dr. Flo. "They scare people because they're also misunderstood."

"Like witches?" said Danni.

"Exactly. Ghosts are often harmless guides that have some kind of *unfinished business*. They are just messengers. Anyway, that's not really why you came today, is it? To talk about witches and ghosts?"

Check It Out!

"Not really," said Wyatt. He thought that maybe Dr. Flo was trying to tell him something, but didn't want to be distracted from his real reason for coming. "Like I told you on the phone, we need help."

"So, tell me," Dr. Flo began as the kids sank into her comfy couch under the portrait of Uncle Clayt. A plate of her oatmeal chocolate chip cookies sat on the coffee table. "How is school going?"

"We don't want to talk about school," Wyatt said, biting into a cookie. "We want to talk about figuring out the woodblock mystery and some other stuff."

"First things first," Dr. Flo said. "Tell me about how school is going, what you're studying, who likes who . . ."

Danni giggled and shared what a nice, but strict, teacher Mrs. Bibbo was, and the song her friend Glenn sang about her on their walk to the Brandywine for "No Kids Left Inside Day." Instead of singing about a farmer and his dog Bingo, he made some substitutions.

"And what did you do at the creek?" Dr. Flo asked.

"We collected leaves from the trees in the park. Then we classified them."

Try It!

She described the many pollution signs posted along the Brandywine and how Mrs. Bibbo suggested that maybe the pollution had something to do with their woodblock. Wyatt then opened up about what was happening at his school. He loved switching classes every forty-five minutes so he didn't have to be in one class for too long. And he liked math and science, but strategic reading and social studies were so boring.

"Except for our field trip to Hagley," he added. "That was pretty good."

Dr. Flo asked a few questions about Hagley, and Wyatt described Docent Don and the gunpowder demonstration as well as the dam, the millrace, waterpower, and milling. "And I got to be a Bancroft," he said.

"Well, that makes sense, since you live in the old Bancroft mill. Didn't I tell you everything is connected? Now, did either of you get any new clues to your mystery from your field trips?"

Wyatt related what Docent Don had told them about the mills of the 1800s sending pollutants, including raw sewage, right into the Brandywine. "Like a big toilet," he said. He described how the walnut from the stickball game disappeared after being caught and submerged by debris in the millrace. He didn't mention the strange man. *Not yet,* he thought.

"I think you may have actually come up with some valuable information for your mystery," Dr. Flo said. "So let's get to work."

Dr. Flo had Danni post a flag on the large wall map where Warner Elementary School was located. Then Wyatt found Hagley and placed a flag there.

"So, what do we know so far?" Dr. Flo said.

Wyatt took the woodblock from his pocket and unfolded the paper rubbings. He read the words out loud for about the hundredth time.

~~Gladness~~
Sadness
Badness
Who will help to stop the Madness?
Y
ch44

Wyatt and Danni then took turns reviewing the clues they had so far: the tennis ball must have come from upstream; it had been in the water for a long time; macroinvertebrates are indicators of water quality; in Wilmington, the Brandywine is posted as polluted; the Christina River is too polluted to eat the fish caught there; there were lots of mills along the Brandywine, many of which added pollutants to the creek; what happens on land can effect water; there was pollution added to the creek in the 1800s and in the 1900s.

"Maybe the madness was the pollution from the mills," Danni said.

"Good thought, Danni," Dr. Flo said. "When would that have been?"

Wyatt thought for a minute. "Probably the eighteen hundreds."

"Where is the tennis ball?" Dr. Flo asked suddenly.

"Actually," Wyatt said, "it's in my sock drawer at home. My mom threw it out, but Benny pulled it from the trash, and I hid it in my drawer because I thought we might need it."

"Did you wash it off before putting it in the drawer?" Dr. Flo asked.

Wyatt wrinkled his nose and shook his head.

"Wash off the tennis ball and bring it next time so we can take a look at it. Maybe it can give us more clues."

"Like, did they even have tennis balls in the eighteen hundreds?" Danni asked.

~ ~ ~ ~ ~

Lunch was a team effort. Dr. Flo put her "Hits from the 50s" on the stereo. They made peanut butter sandwiches to Jerry Lee Lewis's *Great Balls of Fire*, Chubby Checker's *The Twist* and Elvis Presley's *Hound Dog*. For old-people music, Wyatt thought it was pretty good. But when the track got Dr. Flo singing and dancing with David Seville, he thought there were some things about Dr. Flo he needed to know.

"Ooo eee ooo ah ah ting tang walla walla bing bang," she sang. *"Ooo eee ooo ah ah ting tang walla walla bing bang."*

Wyatt looked at the album cover for the name of the song Dr. Flo seemed to enjoy so much. *Witch Doctor*, he read. *Of course.*

Check It Out!

"So, you think the madness has something to do with pollution," Dr. Flo summarized as Wyatt and Danni stood in front of the massive Brandywine Watershed wall map. "How do you suppose you can find out more, seeing as pollution might have come from the land and from upstream? I mean, you two can go check it all out yourselves, but you do have to go to school some days," she added. "Even though you get a lot of days off from school, they aren't frequent enough to rely on for the size of this mission."

"Well," Danni said, "my mom works at Chadds Ford Elementary School and that's right on the Brandywine." She pointed to a spot on the map where the small tributary of Ring Run flowed into the Brandywine right off of Route 1 in Pennsylvania. "But it's in Pennsylvania," Danni said. "Does that make a difference?"

Map It!

"Good question, Danni," said Dr. Flo. "Do you think water, or, for that matter, pollution, cares what state it's in?"

"No," said Danni. "I don't think water or pollution has feelings at all."

"I think you're right about that," Dr. Flo said. "What does your mother teach?"

"Fourth grade."

"Do you think she and her students would be willing to help us out?"

Danni thought for a minute. "I don't know, but if my teacher asked me to help solve a mystery, I'd want to."

Wyatt nodded in agreement. "Especially if it means getting out of school," he added.

~ ~ ~ ~ ~

The sky was fading to dusk by the time Meg Nystrom came to pick up Wyatt and Danni from Dr. Flo's. In the car, a quiet settled in. Wyatt thought about their assignment as he held a copy of the letter and the data worksheet in his hand. He and Danni would do their test site together before approaching his Aunt June to help them. He reread the letter that they had spent the afternoon writing at Dr. Flo's:

"Dear Creek Pals:

You probably don't know us, but Danni is a fifth grader at Warner Elementary School in Wilmington, Delaware and I am in sixth grade at A. I. du Pont Middle School, also in Wilmington. We are hoping you can help us solve a mystery. My dog Benny found a tennis ball in the Brandywine Creek with a message in it. My mom's teacher, Dr. Flo, is a professor at the University of Delaware and teaches Environmental Biology. She's helping us solve the mystery.

We need your help. The message that my dog Benny found (who is a black lab, by the way) goes like this:

~~Gladness~~
Sadness
Badness
Who will help to stop the Madness?
Y
ch44

Will you help us? Since we think the mystery has something to do with water pollution, we are wondering if somebody at your school could explore the nearby creeks and send us information on the form that is included with this letter. If you know other schools near creeks that go into the Brandywine, you can copy and send this letter to them. Then we can all help to solve this mystery.

Here's what we want you to do. Go out to the creek and fill out the data worksheet we included with this letter. Then send it to: Dr. Flo Waters, University of Delaware, Wilmington, DE 19717. Dr. Flo says that you can do the creek study any time of year, but it's really cold in the winter!

Also, we need to know if there is something interesting near your test site that might affect the creek. Like Wilson Run is where Winterthur Gardens is, where there are lots of trees and not many houses or stores. That's important because the land around the water affects the water. So if there's a factory or a golf course or something, we want to know. We call these places "Points of Interest," so if you have them, tell us.

Thanks a lot for helping us. When we figure out this mystery, we will be sure to get back to anybody who helped and tell them what it's all about.

Sincerely,
Wyatt and Danni Nystrom (we are cousins)

P.S. We included the data worksheet and instructions to help you figure out how to study the water. If you don't understand it, contact your high school's science department to help you or email Dr. Flo at Drflobugs@ udel.edu.

P.P.S. Thanks a lot. We really, really, really need your help!

Wyatt never thought of himself as a good writer, but he thought this letter would do the trick. He checked to make sure that the Biotic Index chart, with its pictures of stream critters, was included along with the procedures page. Then he closed his eyes as they drove to Danni's house, hoping that kids from other schools would be as intrigued by the mystery as he, Danni and Dr. Flo were.

Check It Out!

October

Chapter 8

BRANDYWINE BEAUTY MARK

Dr. Flo led Wyatt and Danni to the stream and explained that the morning's agenda was water-quality testing. If they were going to ask kids from other schools to help solve the mystery by doing the tests, then they needed to know how to do the tests themselves.

A pair of kayakers floated through the small rapids of Brandywine Creek State Park. Behind them, tulip trees, with their waxy yellow leaves and upright clusters of seeds, stood straight and tall. Wyatt watched the current carry the kayakers downstream toward the takeout ramp, close to where his mother, father, aunt, and uncle chatted at a picnic table. The adults had agreed to wait for the kids to complete their water-quality research with Dr. Flo before going on their family hike. Benny lay on the ground beside them, whining as he watched Wyatt and Danni upstream.

Map It!

Danni spread the data worksheet and identification papers on the ground along the creek and placed a rock on each to keep them from blowing away. The data sheet was smudged with mud, giving it the quality of an authentic field investigation.

"First, we'll measure the physical aspects of the creek, including depth, width, velocity, turbidity, and temperature," Dr. Flo said, unpacking equipment. "Moving, clear, and cold is how we want our creek water, so always choose an area with rocks on the bottom and bubbling water, a *riffles* area."

The cousins waded into the creek. It took several seconds for the water to seep through the canvas of Wyatt's old sneakers. "Well, at least we know the water's cold," he said.

Even with her tall rain boots blocking direct contact with the water, Danni could feel it too. "Can't we just test from the shore?" she said.

They measured the temperature, turbidity, depth, and width of the creek. With a cork and a stopwatch, they determined the velocity of the stream. Traveling at the rate of one meter per second, the water was moving pretty quickly. *Which would explain why the kayakers are here*, thought Wyatt. Back on the stream bank, they recorded their results on the data sheet.

Try It!

"We want all of the tests to be taken in areas of similar velocity," Dr. Flo said, "so that the sites are comparable. It wouldn't be fair to compare an area of fast-moving water to an area of sluggish water, now, would it?"

"I don't think so, but I'm not sure why," Wyatt said.

"You'll see soon enough," said Dr. Flo.

Danni and Wyatt started the chemical tests, following the directions in the manual. They gathered water samples in small glass bottles to test pH and dissolved oxygen. They added and mixed various chemicals and quickly assessed the pH of the creek to be *seven*, neutral.

"Just what we want," Dr. Flo said. "The critters we're looking for can only live in water with a pH of 6.5 to 8.5. Otherwise, the water is too basic or too acidic to support aquatic life."

Check It Out!

The dissolved oxygen test was far more involved and resulted in various transformations of the water from clear to gooey brown to yellow and then back to clear again. The result was a *ten*. "Ten *what*?" Dr. Flo asked. "Unlike pH, you can't just say *ten* without giving units of what you are measuring. In the case of dissolved oxygen, you are measuring parts per million. Let's say you have a million pieces of water. Ten of them would be oxygen. A value of ten or higher is very good, and provides ample oxygen for most critters that need it, even the picky ones like trout and some of the macroinvertebrates that we'll be looking for."

Check It Out!

Wyatt looked at her. "So that's why we have to pick a rocky area with bubbly water."

"The macroinvertebrates need good oxygen levels," Dr. Flo said. "Otherwise, we can't compare the results from site to site fairly. Rocks provide shelter, and the water bubbling over them helps to provide oxygen. And speaking of macroinvertebrates, now comes the best part of the research."

"Why do we just look for macroinvertebrates?" Danni asked. "Why don't we look for salamanders or fish or frogs?"

Dr. Flo thought for a moment before answering. "What can salamanders or fish or frogs do, if the water gets impaired, that the macroinvertebrates can't do?"

"Leave," said Wyatt.

"Exactly. So we look at the critters that can't just leave if the going gets tough and the water gets dirty. We look at the macroinvertebrates as *indicators* of water quality."

Danni and Wyatt filled several empty yogurt cups with water and placed them on the bank of the creek while Dr. Flo unrolled the square meter seine net. She laid it alongside several paintbrushes and a rectangular plastic tub Wyatt's dad had given him, salvaged from a seafood store. When Luke Nystrom had appeared in the doorway of the condo a few nights earlier and presented the tub as a gift, Wyatt gagged at the fishy stench. But when Luke explained its purpose, Wyatt volunteered to clean it out. Now filled halfway with water, it made the perfect temporary home for the macroinvertebrates they were hoping to find.

Wyatt, Danni and Dr. Flo waded back out into the creek to the riffles area. As Dr. Flo held the seine net in place, Wyatt and Danni picked up rocks in the one-meter area upstream from it. They scrubbed each rock, dislodging any critters so they would float into the net, and then returned all but one rock to the stream bottom. After covering the entire one-meter area, Dr. Flo and Wyatt carried the net back to the stream bank, making sure not to spill anything off of it. Danni brought the rock and placed it in the tub.

Wyatt pulled off his sneakers, exposing his shriveled white toes. Danni dumped the water from her boots. "Did you check to see if you have any macros in there before you dumped them out?" Wyatt joked.

"Look," Danni said, ignoring Wyatt. She pointed to a small insect on the net. Carefully, she removed it with the paintbrush and placed it into a yogurt cup. Though it lay flat against the bottom of the cup, small fanlike projections along its abdomen fluttered. "What is it?"

Wyatt looked to Dr. Flo. Instead of answering, she handed him a set of laminated cards with photos and information about macroinvertebrates. "See what you come up with," she said. "I'm going to join your parents."

Check It Out!

"How many tails does it have?" Wyatt asked Danni.

"Three."

"It must be this." Wyatt showed Danni a photo of the mayfly with its three tails and abdominal gills. "A mayfly. Remember we found these in Alapocas Run?"

Danni checked for the mayfly on her identification chart. "Mayfly larva, Class One. Pollution sensitive; found in good-quality water." She checked off the space below the mayfly on the data sheet. "That's a good one."

Wyatt lifted the rock from the tub. Small clusters of sand grains crusted its bottom. Peering closely, he detected a tiny orangish head poking out from one of the clusters. "Caddisflies, like the ones we found in Alapocas Run," he noted. "This rock is covered with their shelters."

Danni looked at the chart and checked off *caddisfly*. "Another good one. Class One."

Returning the rock, Wyatt spotted what looked like a scab creeping almost imperceptibly along the bottom of the container. He picked it up and turned it over to observe a set of six delicate legs translucent against the dark body. "A water penny," he identified, flipping through the photo cards. He placed it on his cheek. "Or a beauty mark," he laughed.

"Yuck, get that off your face!" Danni said, though she had to admit, it certainly did look like a beauty mark. She checked off the water penny, another Class One organism, and continued the assessment as

Wyatt picked critters off of the net, identified them and placed them in the yogurt cups: small freshwater clams, snails, scuds that resembled tiny shrimp, mayflies, caddisflies, and water pennies.

"OK," Danni said. "Let's see what we've got. Mayfly, caddisfly, and water pennies are all Class One, found only in good-quality water. We got scuds and clams from Class Two for fair-quality water, and these worms from Class Three can be found in any quality of water." Using the formula Dr. Flo had shown them, Danni did the calculations in her head to determine the Biotic Index. *Two points for each type in Class One; one point for each type in Class Two; zero points for each type in Class Three.* "Eight points," she said.

Wyatt looked at the chart that explained what the Biotic Index means in terms of water quality:

0 = very poor water quality
1-6 =poor-fair water quality
7-9 = good water quality
> 10 = excellent water quality

"Good, but not excellent," he said, though he remembered Dr. Flo's words of caution. "Good by Mother Nature's standards, not ours. It doesn't mean you can drink water straight from the creek," she had told them.

Check It Out!

"Pretty good for our first time," Danni said as she filled in the data sheet.

Wyatt returned the rock to the spot where they had gotten it, recalling Dr. Flo's words, "If someone removed your home, wouldn't you want them to return it when they were done?" They washed out the net and gathered up the equipment, excited to tell Dr. Flo their results. *We may not have solved the mystery,* Wyatt thought, *but that was pretty cool.*

~ ~ ~ ~ ~

The hike through Tulip Tree Woods was slow-going for the adults. The rough, uneven terrain kept them to a stroll. Wyatt, Danni, and Benny ran ahead to one of the intersecting blue-gray rock walls.

"C'mon," Danni called, springing to the top of the low wall.

Wyatt followed her lead as Benny sniffed enthusiastically at anything he could reach from his leash. At the fork in the path, they jumped down from the wall to follow the Indian Springs Trail. Leaves crunched underfoot. Nuthatches and chickadees flitted from tree to tree. Nearby, a squirrel squawked a warning, then zipped up a poplar tree and popped into a hole the size of a small pancake. Wyatt continued his gaze up from the hole to the treetops. "Check it out," he said.

Danni looked up. All around them were tulip poplars, some standing tall, straight and limbless, others deformed and twisted. A slight breeze picked up, sending yellow leaves fluttering down as quietly as snow. In the distance, they heard bits of adult chatter. Wyatt, Danni, and Benny scampered down the trail toward the creek, keeping some distance between them and adult supervision.

At the creek beach, the children watched Benny strut straight into the water and lay down.

"That is so cold," said Wyatt. "How does he do that?"

Benny hoisted himself up and shook, spraying the kids.

"Benny," Danni whined, but quieted when the dog suddenly froze. A low growl rose from his throat and erupted into a menacing bark.

Wyatt and Danni looked to see a pale-faced boy staring at them from behind a tree. "Hi," Wyatt called tentatively. Benny quieted. The boy didn't respond, but stepped out from behind the tree. His dark knickers, buttoned black boots, and cape looked peculiar and out of place. The boy turned and walked slowly in the other direction. Placing his cap on his slicked hair, he disappeared around a bend in the trail.

"There you are," June declared as she, Luke, Meg, and John Nystrom approached from the other direction. "Couldn't wait for the old folks, could you?"

Wyatt placed his forefinger over his lip, signaling Danni to keep quiet, and shook his head. "Nope," he replied. "At the rate you move, you miss all the good stuff."

"What did we miss?" asked Luke.

"Nothing," Wyatt responded. "I was just kidding."

"What was Benny barking at?" asked Meg.

Wyatt glanced at Danni. "Oh, you know, he just gets spooked sometimes by the ghosts and lions and tigers . . ."

"And bears. Oh, my," the adults said together, laughing as they joined Wyatt in his little joke.

~ ~ ~ ~ ~

"There's something I need to talk to you about," Wyatt whispered into the phone as he recorded his message on Dr. Flo's answering machine. He had promised Danni that, if they weren't going to tell their parents, he would definitely confide in Dr. Flo about the strange boy. "Please call me back as soon as you can."

~ ~ ~ ~ ~

"C'mon," called the pale boy with the slicked-back hair. "Follow me." Benny barked playfully as he chased the boy, nipping at his shiny boots. Wyatt picked up the boy's cape that had fallen from his shoulders and ran to catch up. The boy stopped. Just beyond the trail, sitting beside a campfire, a man sat, wearing singed, tattered clothing. The man beckoned to the boy, his eyes intense.

"It's not your fault," he said to the boy in a sad voice.

"Keep this," the boy said. He took the cape from Wyatt and handed it to the man. When the man reached to take it, Wyatt saw that his arm was badly burned, oozing a yellow liquid.

Wyatt woke to Benny's wet nose nuzzling his own arm. He stroked the dog's neck. "Weird stuff is going on, Benny," he said.

Field Trip!

Chapter 9

DREAM ON!

Date: October 7
Time: 1pm
Location: Brandywine Creek State Park under bridge
Method of Sampling: seine net sampler
Weather Conditions: warm and sunny, about 75 degrees
Weather in Last 24 Hours: warm and sunny
Biotic Index: 8
Points of Interest: Lots of Woods, near Winterthur
Gardens
Study completed by: Wyatt and Danni Nystrom
pH=7   Dissolved Oxygen = 10 ppm

Danni copied the most essential information on to a red sticker: *BCSP, BI = 8.*

She placed the sticker on the wall map in Dr. Flo's living room. Wyatt sat on the couch under Uncle Clayt's portrait, squeezing the tennis ball.

Sticker It Red!

"Where did you test?" Dr. Flo asked.

"The creek," answered Danni. Dr. Flo gave her a look, prompting Danni to *be more thorough, be more thoughtful.* "Oh, you mean Brandywine Creek State Park," she said. "That's what *BCSP* stands for."

"Yes, and what is BI?"

"Biotic Index," she said.

"Precisely," said Dr. Flo. "And since the Biotic Index at that site was less than ten, we use a red sticker. If it was more than ten, we would use a blue sticker. Do you remember what the significance of a Biotic Index greater than ten is?"

"Really clean water," Wyatt said.

72

"Right," answered Dr. Flo. "Any area with a Biotic Index less than ten is going to get a red sticker, which means that area of the creek is impaired and not as clean as it could be. So, just by glancing at the map, we can see a snapshot of your results and see which areas are in good condition and which areas are not. If we need to, we can go back to your data sheets for more specific information. Now, where else do you think you should run tests?"

Wyatt slid off the couch to survey the map. "We should probably go back to Brandywine Park, the one by the zoo, even if we know it's polluted," he said. "We should get a Biotic Index."

"Good idea," encouraged Dr. Flo. "Where else?"

Wyatt traced his finger up the Brandywine and stopped at the border between Pennsylvania and Delaware. "Maybe here," he said, where the creek looped at the state line.

Map It!

Danni continued to trace upstream. "Here," she said, designating a spot near Chadds Ford Elementary School. "My mom's school is right here near the Brandywine."

Wyatt and Danni continued following the stream up the map, pointing out possible test sites.

"Now, remember," Dr. Flo cautioned, "You may need help from your Creek Pals. Be sure to keep that in mind. And many of the sites may not be accessible, due to the terrain or because it's private property."

"That's not fair," complained Wyatt. "It's research."

"But it's the law," Dr. Flo explained. "The water is public but, in many places, the land around the water and even *under* the water is private property. And speaking of the law, I have these for you." She pulled two fishing licenses from a drawer. "Just in case," she added, pinning a license on Wyatt's shirt. "It keeps you legal when you're borrowing critters from the creek. I have my educator's permit, but since I won't be with you most of the time, you should carry these with you when you do your testing. Technically, you don't need these until you're sixteen, but it makes you look more official." She winked and pinned the other license on Danni's shirt.

Check It Out!

"Thanks," they said.

"Now," Dr. Flo got serious. "Tell me about the strange goings on you've been witness to."

Wyatt began by confessing that he hadn't told anyone about his visions of the damaged man at Hagley. But when the same man appeared in his dream, along with the strange, old-fashioned child that both he and Danni saw in the park, he decided to come clean.

"Dreams come from your heart as well as from your head," said Dr. Flo. "It's important to listen to them, for they may tap into something that may be important for you to understand. What exactly happened in the dream?"

Wyatt reviewed the dream in as much detail as he could remember and Danni described what they had both seen in the park. "It must be real," she added, "'because Benny saw the boy, too."

"Animals often sense more than we do," Dr. Flo said. "Were either of you scared at any point during those sightings?"

"No," said Danni.

"They were weird, but they just made me curious, really," added Wyatt.

"Something in your heart must have put these two very different people into one vision for you, Wyatt. Let's see if we can piece this together."

Everything is connected, Wyatt thought.

~ ~ ~ ~

"The internet makes things so simple to research," said Dr. Flo. They sat in her study, ready to research history of the Hagley Museum and Brandywine Creek State Park. "Imagine back in my day, when we had to go to the library and search the archives for actual journals and newspaper clippings for information." The children looked at her in disbelief. "Old as dirt, I am," she added as she typed into the search line: "Hagley accidents, fires, explosions." Numerous links popped up, but one in particular caught Wyatt's eye: "Explosions at the DuPont Powder Mills."

Check It Out!

"Check that one," he said.

Just as Docent Don had described during Wyatt's field trip, deadly explosions took place periodically over the hundred and seventeen years of black powder production at the DuPont mill. "Five Explosions with Largest Number of Deaths," Wyatt read aloud. A list of explosion dates followed. Dr. Flo's eyes grew big, but she remained quiet as Wyatt and Danni read the list of dates followed by the number of casualties.

Next, they searched the history of Brandywine Creek State Park and discovered that it had been a former livestock farm belonging to Henry Francis du Pont. Born in 1880, H. F. du Pont grew up on the Winterthur estate where his home, gardens and furniture were now open to the public as a museum. He purchased the present Brandywine Creek State Park in segments to put some distance between his home and "those stinkin' cows," Danni interjected, interpreting the more diplomatic language used on the website. He had the boundaries of each purchase outlined with rock walls.

"Wasn't it a du Pont who started the Hagley gunpowder mill?" asked Wyatt. "Is this a different du Pont?"

Searching the genealogy of the du Pont family, they discovered that Henry Frances du Pont was one of dozens of great-grandchildren of the gunpowder mill founder, Eleuthere Irenee du Pont, also known as E. I. or Irenee.

"Like there aren't enough du Ponts to keep track of, and this one goes by three different names," muttered Dr. Flo.

Check It Out!

"Let's look back at the explosion dates and see if anything stands out," suggested Dr. Flo. She returned to the link to the five deadliest explosions at the gunpowder mill:

3/19/1818: 34 deaths
11/30/1915: 30 deaths
4/14/1847: 18 deaths
2/25/1863: 13 deaths
10/7/1890: 12 deaths

"You know, at Hagley, there was a poem at the schoolhouse about twelve people dying," Wyatt said.

Danni gasped. "Look at the date when there were twelve deaths. October 7. That's the same day we did the stream study at Brandywine Creek State Park."

"Let's get some fresh air," Dr. Flo suggested.

They left the computer and went through the living room. On the way to the front door, Wyatt looked at the portrait hanging over the couch.

"Was he smiling before?" he asked.

"My Uncle Clayt? He always smiled," said Dr. Flo.

Danni looked at Wyatt sternly. "Don't be making things even weirder than they already are, Wy."

~ ~ ~ ~ ~

Dr. Flo rocked gently in her wicker chair while Danni and Wyatt swung on the porch swing. "So, do you think you understand your dream now?" she asked them.

Danni and Wyatt pieced the clues together like a puzzle. The singed man must have been killed in the disastrous explosion of 1890 at the DuPont gunpowder mills. That would explain his unsightly appearance. At the time of the explosion, in 1890, Henry Francis du Pont was ten years old, so the outfit he wore in the park and in Wyatt's dream made sense. As a member of the wealthy family who owned and ran the powder mill, he must have felt terrible about the destruction and the loss of lives in the Hagley community, though he could not possibly have been responsible, being only a child at the time. Maybe Henry's legacy of Winterthur and Brandywine Creek State Park were, in part, a sharing of the fortune of the du Pont wealth that had been created at the expense and sacrifice of ordinary people.

"Wow," said Wyatt, tossing the tennis ball back and forth between his hands. "All those explosions; maybe that's the *madness*."

"But what about the tennis ball? About whether it could have been from the eighteen hundreds?" said Danni. "That would tell us if this is the madness."

Dr. Flo intervened. "I think we've done enough work for today, don't you think? Besides, we have some worms to feed. Come on, help me take the compost out to the worm box, will you?"

"I guess you'll just have to wait," Wyatt said to the tennis ball. He looked at the faded USLTA stamp.

Chapter 10

GETTING TO THE ART OF IT

A great blue heron glided over the creek. Wyatt and Danni and the seven other Saturday art class students left the cobblestone walkway of the Brandywine River Museum and headed toward the water. A guest for the day, Wyatt followed behind Danni and her new friend Eliza. Like the two Saturdays before, the students settled on the benches facing the creek. Grinning at them was a life-size, bronze sculpture of a curly-tailed pig. Across the creek, shafts of sunlight filtered through the leafy trees. In silence, the students looked around at the angles of the trees, the flaming reds and yellows of the maples and poplars, the coarse grain of the grinding wheel, the rusted bridge crossing the creek. Catching these varied shapes, textures, colors, and a *sense* of this place and putting it onto paper was the object of the class. But Danni just wanted to paint.

Map It!

"Time," called Tina, the art teacher. Taking the cue to end their *meditation*, the students jumped from their benches and dashed up the River Walk, stopping momentarily to point out a track print or climb the sycamore tree frozen in its mid-fall toward the creek. They darted under the Route 1 Bridge, splashed through small streamlets, and raced through the open field between massive utility posts. They stopped at the boardwalk to wait for Tina, who ambled along with her bag of drawing materials.

"What is that?" Eliza asked. She pointed to a small dropping on the deck.

Tina bent to pick up the object.

"Ick," cried the kids. "Isn't it . . . ?"

Tina pulled the dark, moist dropping apart, revealing purple berries and some fur. She broke into a rap.

"It starts with an 'S',
ends with a 'T,'
comes outta you,
and it comes outta me."

The kids stared at her. *Is this really how teachers act on weekends?* Wyatt thought. Tina continued:

"I know what you're thinking,
don't call it that,
be scientific
and call it SCAT."

Rap It!

"From a fox," she added matter-of-factly. "Who clearly eats his fruit and veggies with his meat."

Check It Out!

Danni and Wyatt looked at each other. "Another member of the Dr. Flo club," whispered Danni. Wyatt nodded in agreement. "We'll explain later," she said to Eliza.

The students settled along the boardwalk, sketchpads and pencils in hand.

"I'm going to keep walking," Wyatt told Danni. "I'll be back in a few minutes. I'll be too bored if I just stay here watching you draw."

Wyatt followed the meandering boardwalk through the wetlands. A flash of red caught his eye. It disappeared as the red-winged blackbird landed on a cattail and folded in his wings, then called out in a liquid trill. Bent from the weight of the bird, the cattail released a collection of fluffy seeds that parachuted away silently. Wyatt stood still, observing, the sun warm on his face. Moments passed. A black snake slithered silently below the boardwalk. A flock of Canada goose honked above as they adjusted their flight into the shape of a checkmark. A spider scurried across her zigzag web.

Check It Out!

Wyatt resumed his stroll to the end of the boardwalk. A burst of twittering from above distracted him. He looked up to see, perched on one of the utility posts, a bald eagle squeaking as his mate tore the flesh from a fish.

Running back along the boardwalk, he called to Danni, "Eagles, eagles!"

Danni turned from her sketch. She looked up to see one of the eagles, head and tail polished white, leap from its perch and glide silently up the creek.

"A great sign," said Tina. "It means a healthy habitat. Of course, it's not surprising since we're in a wetlands area: very important for keeping the stream clean. Those eagles are such a majestic sight. Now think. What did you just see? But adjectives only." She nodded as the children called out words.

"White."

"Big."

"Brown."

"Regal," called out one girl.

"Ah, now that's a powerful image. Give me more." Each student attempted to top the last one's description.

"Sharp."

"Massive."

"Soaring."

"Tough."

"Now," she continued. "Which is more descriptive to you? 'Brown' or 'Regal?' 'Big' or 'Massive?' That's what we're trying to do with our art: to take the obvious, like 'Brown' or 'Big' and bring it to life. Translate your words and emotions into your art." She pulled out a large book of paintings and flipped to one of a pig. "Take a look at how majestic and alive everyday sights can be," she said. "Isn't this pig grand? Look how you can almost feel how coarse the pig's hair is and how moist his snout is. And think about how he makes you feel."

"He looks like he's smiling," Eliza said. "He's so cute. I just want to go up and hug him." The kids giggled at Eliza's enthusiasm, but nodded in agreement.

"Now," Tina continued as she flipped to another painting, "look at the difference between Jamie Wyeth's *Portrait of a Pig* and his grandfather N. C. Wyeth's illustration of *Robin Hood*."

Danni looked carefully. She could tell that different people had created the paintings, but she couldn't determine what made them different. "Well, I don't want to hug Robin Hood," she said.

"Yes, you feel differently about this painting. Maybe a little more fearful. Now, look at one of Andrew Wyeth's most famous paintings." She turned to a page with a picture of a woman in a field dragging herself toward a home. The caption below read *Christina's World*.

"Andrew was N. C.'s son and Jamie's father." She flipped back through the examples. "Three generations of artists with three different styles, all having practiced their art right here in the Chadds Ford area. Just like you. Your job, whether in this class, another art class, or life itself, is to create your own style as you bring the everyday alive. Remember that. Now, let's pack up and head back. We'll continue next week."

Check It Out!

~ ~ ~ ~ ~

Eliza's dad waved to the three kids, motioning for them to get in the car. With all three in the back seat, he turned out from the parking lot onto Route 1. He had told Eliza that she could bring a friend or two on their family trip to Longwood Gardens that afternoon, and Eliza had asked Danni, her newest friend from art class. Since Danni lived in Delaware, Saturdays were the only time they could spend together. And since Danni had told Eliza about the mystery she and Wyatt were working on and how they had unraveled some history involving some of the du Ponts, Eliza invited Wyatt to join them, too. After all, it was well known that Longwood was another du Pont legacy.

"What's a conservancy?" Wyatt asked Eliza's dad as they passed the sign designating the Brandywine River Museum as a project of the Brandywine Conservancy.

"It's an organization created to protect or conserve something," Mr. Shaps answered. "Usually a natural area or a wildlife area. The

Brandywine Conservancy protects things related to the Brandywine Valley, including the environment and the art, since so many of the local artists were inspired by the environment."

Wyatt thought about the eagles and how, if he were an artist, he would draw the female shredding the fish. He would call his picture "The Circle of Life." Or maybe, if he thought about the fish's point of view, he would call it "The Circle of Death."

"Are wetlands a part of the conservancy? Because the art teacher said something about a wetlands helping the environment," Wyatt said. "I know they're not just full of mosquitoes, but how do they help?"

"Wetlands act like a sponge," Mr. Shaps said. "They can absorb pollutants from dirty water and release the clean water back into the ground or the stream. And they provide great habitat."

Check It Out!

Wyatt thought of the variety of life he saw in the wetlands during the short time he watched from the boardwalk: the spider, the snake, the cattails, the different birds. He liked having adults around who could answer all of his science questions.

"So," Mr. Shaps said when Wyatt didn't say anything for a few minutes. "We'll go back home, have some lunch, and then head out to Longwood Gardens."

"There's my mom's school," interrupted Danni. She pointed to Chadds Ford Elementary School. "I wanted to take art classes, and my mom figured she could work at school while I took classes here. Oh, yeah," she interrupted herself. "Mr. Shaps, she wanted to know if she could just meet us at Longwood instead of coming to your house. Is that OK?"

"Sure, Danni. Let her know that we'll probably be leaving Longwood at about four o'clock. Can you text her?"

"See why I should have a cell phone, Dad?" Eliza said as Danni texted her mom.

"There's *my* school." Eliza pointed to the low, gray building surrounded by rolling, open fields. A large sign read HILLENDALE ELEMENTARY SCHOOL.

Map It!

"Oh, I get it," Danni said. "Hill and dale. A dale is like a valley, right?"

"You've got some smart friends, Eliza," Mr. Shaps said as he drove into their neighborhood. "Well, we're almost home."

The neighborhood reminded Wyatt of something. It was densely treed and had a campy feel, so unlike his neighborhood. The road dipped down, crossed a small creek, and climbed back up to Turkey Hollow Road.

"Are there turkeys here?" Wyatt asked.

"There haven't been turkeys here since this neighborhood was built. That's what they say about developments," Mr. Shaps said. "They're often named for what disappears when they're built."

"That's sad," said Wyatt, thinking that it would be pretty cool to see a real turkey. The only ones he ever saw were on the Thanksgiving table. On a round plate. *The Circle of Death*, he thought. "And what's that little creek we passed over? Do you know where it goes?"

"I'm not really sure," Mr. Shaps said. He pulled into the driveway. "All right, guys, go kill some time while I make lunch. Is grilled cheese good for everyone?"

"Anything but turkey sandwiches," said Wyatt.

~ ~ ~ ~ ~

The kitchen table was set with plates of grilled cheese and bowls of tomato soup. "It's my favorite lunch," Eliza told her friends. Mrs. Shaps joined them as they sat down to eat while Mr. Shaps unfolded a map at the kitchen island.

"You want to take a look, Wyatt?" Mr. Shaps pointed to the intersection of Woodchuck Way and Turkey Hollow Road. "Here's where we are, and here's the creek."

Map It!

The origin of the creek, the *headwaters* as Mr. Shaps described it, was right in the neighborhood. "That's where the water table meets the land surface."

"He's a science guy," Eliza said, apologetically. "Dad, we don't get it."

"It's where the water in the ground, known as groundwater, reaches the surface of the land and seeps out to start a creek."

"Don't creeks come from the rain?" asked Wyatt.

"Great question, Wyatt. But no, most creeks are fed water from the ground. If creeks were fed by rain only, we'd have dry creeks during any periods of no rain."

Wyatt thought for a minute. Of course creeks don't dry up whenever it doesn't rain. So the water in a creek had to come from somewhere else.

"See all of these little creeks in the area?" Mr. Shaps pointed to the nearby creek, Ring Run, and a few other creeks just south of the neighborhood. One at a time, they traced the path of each creek on the map from its source to a merge. Several of the creeks headed south to the Red Clay Creek. Several headed east, merging with the Brandywine.

Map It!

"Danni, Eliza's creek goes right by your mom's school," Wyatt noted. "I have an idea. Mr. Shaps, will we have a few minutes after lunch before we go to Longwood?"

"Sure. But eat your soup; it's getting cold."

~ ~ ~ ~ ~

Through the woods behind her house, Eliza led her friends down a hill to the stream. "I never thought about where this water goes," she said. "I just like to explore here. Look." She picked up a white quartz rock from the stream. Several water striders that had been skimming along the water's surface scattered. A crayfish disappeared into a puff of cloudy water.

Wyatt took the tennis ball Mr. Shaps had loaned him. "You may never see it again," Wyatt had warned. Mr. Shaps said that he didn't need it back; it was an older ball that didn't have much bounce left and wouldn't be good enough to use on the court again. If Wyatt had a good use for it, then by all means, he should take it.

Using the permanent marker Eliza had brought, he wrote on the yellow ball just above the USTA insignia: *10/21, property of Wyatt Nystrom.* When Wyatt asked what the initials stood for, Mr. Shaps had answered that it was the *United States Tennis Association.* Wyatt could have sworn that the torn white tennis ball at home had different lettering in its insignia, but maybe he was wrong. He placed the ball in the creek and watched as the water carried it a short way before it got hung up behind some rocks. He pushed it with his foot and watched as it journeyed slowly downstream.

~ ~ ~ ~ ~

A kaleidoscope of chrysanthemums and enormous, odd-shaped gourds and squashes bordered the walkway at Longwood Gardens. Wyatt, Danni, and Eliza ventured as far as the largest of the garden's three lakes. Mr. and Mrs. Shaps had allowed them to explore the grounds on their own until three o'clock, when they would regroup at the Peirce-du Pont House.

Map It!

The air was crisp and the sky provided a brilliant blue backdrop to the scarlets, oranges, and golds exploding from the surrounding flora. Wyatt and the girls ran along the pathway bordering the lake and around the Italian Water Gardens. They followed the bridge over Hourglass Lake and walked through the wooded path, bordered by ivy, bottlebrush buckeye, and mountain laurel. The shouts of small children caught their attention. They followed the voices farther into the woods where they found a spectacular tree house hosting a dozen kids and their parents. They climbed the wooden staircase of "The Birdhouse" and took a quick look out the window. Several crows cawed from the surrounding trees. The trio dashed back down the steps and continued to the meeting place where Mr. and Mrs. Shaps chatted with a Longwood Gardens guide.

"Did you have fun?" asked Mrs. Shaps as the kids plopped down on a bench.

"Yeah, it's really cool here," said Wyatt.

"My favorite was the treehouse," added Danni.

"Ah, mine too," the guide agreed.

"Let's head over to the Conservatory," suggested Mr. Shaps. "Mrs. Shaps and I already toured the Peirce-du Pont house. We figured you kids wouldn't be interested in it."

"Which du Pont is that?" Wyatt asked. "Is he related to the guy who started Hagley? Or the guy who lived at Winterthur and started Brandywine Creek State Park?"

Mr. Shaps looked at Mrs. Shaps with an "I told you so" look. He had told his wife earlier about how inquisitive and thoughtful Wyatt was.

"Actually, yes, they are related," said the guide. "Pierre du Pont was a great-grandson of E. I. du Pont, 'the guy who started Hagley.' Henry Frances du Pont, to whom we attribute Winterthur and Brandywine Creek State Park, was a great-grandson of E. I. also. Henry and Pierre were second cousins."

Check It Out!

"You've lost me," said Mr. Shaps.

"He's a science guy," explained Mrs. Shaps. Eliza looked at Wyatt and Danni with her own "I told you so" look. "Please, go on," Mrs. Shaps urged.

Wyatt listened, intrigued by the interconnections of one of the most successful families in American history.

The guide continued. "Pierre was a successor of the DuPont Company and he made a fortune. He purchased this property from the Peirce brothers, farmers who planted the original arboretum. Pierre traveled a great deal and had a passion for gardening. He expanded the property into a showcase of gardens as a tribute to his family and the natural world. He constantly added new treasures to his property to surprise his guests, including the Open Air Theatre with secret fountains that showered his unsuspecting nieces and nephews."

Suddenly, classical music filled the air as sprays of water shot up from a set of fountains across the walkway. "Right on cue," the guide said. "Can you imagine, as we enjoy this production, just over three hundred years ago, this property was a wilderness inhabited by Native American Lenapes living off the land as hunters and fishermen."

Wyatt looked amazed. *That is unbelievable,* he thought.

The guide continued. "As Pierre approached the end of his life, he wanted this showcase opened up to the public for everyone to enjoy. We are fortunate indeed. Now, don't stand here listening to me ramble on; go watch the fountain show!"

~ ~ ~ ~ ~

June Nystrom was waiting outside the visitor center when Wyatt, Danni, Eliza, and her parents emerged. She got out of the car. "Thanks so much for having the kids," June said as Danni and Wyatt climbed into the back seat.

"It was our pleasure," said Mr. Shaps. "They seemed to really enjoy it. And I'll tell you. Wyatt, he's your nephew, right? He is quite the curious one. He knows more about this area than I do."

"That's because he's a science guy," Wyatt whispered. Danni giggled.

June drove out of the parking lot and turned on Route 52. "Drat," she muttered within seconds of turning. "I wanted Route 52 South, not North." She drove a short way down the narrow road and pulled over onto the shoulder to check her map.

"Look at that," Danni said, pointing to a stone monument just beyond the car. "Can we check it out?"

"Sure," said June. "I'll just quickly reroute our return trip home since there doesn't seem to be a place to turn around."

Map It!

~ ~ ~ ~ ~

Danni and Wyatt stood in front of the monument and read the inscription on the bronze tablet.

INDIAN HANNAH
1730-1802
THE LAST OF THE INDIANS IN CHESTER COUNTY
WAS BORN IN THE VALE ABOUT 300 YARDS TO THE EAST ON THE LAND OF THE
PROTECTOR OF HER PEOPLE THE QUAKER ASSEMBLYMAN WILLIAM WEBB
HER MOTHER WAS INDIAN SARAH AND HER GRANDMOTHER WAS INDIAN JANE OF THE
UNAMI GROUP
THEIR TOTEM THE TORTOISE OF THE LENNI-LENAPE OR DELAWARE INDIANS
MARKED BY THE PENNSYLVANIA HISTORICAL COMMISSION AND THE
CHESTER COUNTY HISTORICAL SOCIETY
1925

Behind the monument, a chain-link fence separated them from the *vale*, a sloping woodland area that looked like the background of the Lenape diorama Wyatt had seen at the Hagley Museum. He imagined the land in front of him as an encampment of Lenape Indians burning and scraping out the straight trunk of a tulip tree to create a canoe; spearing shad in the nearby Brandywine and hanging it to dry; hauling deer back from the hunt to supply them with meat for food, skins for clothing, and bones and sinew for tools, cordage, toys, and ornaments. He pictured the women pounding corn with mortar and pestle and young boys flint knapping arrowheads. Four crows flew overhead, calling out in raspy voices.

Check It Out!

"All right kids, let's go," called June. "I think we'll take the scenic route."

Wyatt and Danni returned to the car. June drove onto Route 52 North. They turned again and drove along rolling hills. Wyatt wondered if the open fields and housing developments had something to do with why Indian Hannah was *"the last of the Indians in Chester County"*.

Did they die out? he wondered. *Or did they leave? What happened to them?*

~ ~ ~ ~

"Look at those hideous things," June commented as they crossed the Brandywine Creek and headed south on Route 100. She pointed to the utility posts looming overhead.

"Oh, yeah," Danni said. "We saw a pair of bald eagles on one of those earlier today."

"Really? I'll have to have my fourth-graders be on the lookout for them," said June. "I got a chance to go through your 'Creek Pals' materials today and thought I'd take my class out to collect some data for you next week. The kids will love looking for eagles while they're out."

They continued the drive down the road as June pointed out sites of interest: the old gristmill that was home to artist Andrew Wyeth; the driveway of the house featured in the film *Marley and Me* about the world's worst dog; the Chadds Ford Historical Society's Visitor Center. They continued down Route 100, paralleling the Brandywine as it wound its way toward the Delaware state line.

Map It!

June checked the rearview mirror. Danni and Wyatt leaned into each other in the back seat, their eyes closed She smiled silently as they drove past the gates of a property that was once an old Lenape Indian village. The tortoise totems adorning the gates meant home was not far away.

Field Trip!

Chapter 11

CLEAR SKIES

The sun streamed through the window of bus forty-two, warming Wyatt's face. It felt good after the last three days of rain. It was Friday morning, the sky was bright blue, and the air was warm for the end of October. His bus lumbered along roads slick with wet leaves that had fallen off the trees with all the rain.

In gym, Wyatt's class played tennis. At the end of class, Wyatt helped collect the scattered tennis balls. *Sixth-graders can't really volley a tennis ball*, he thought. The balls were all over the place. He checked the insignia on each one and couldn't help but be bothered. He was sure that the USTA label was different from what was on the ball Benny had found. He promised himself to check the insignia when he got home.

~ ~ ~ ~ ~

When Wyatt arrived at home, he saw several missed calls from Danni. Her voice on the message was excited. "Call me as soon as you get this."

"Mom took her class out today and tested the creek, and guess what," Danni said when Wyatt called. "She found your tennis ball. The one you put in the creek at Eliza's house. Remember? It must have been from all that rain. One of the kids in her class found it and was tossing it around. Mom told him to stop fooling around and get back to work. Then she took the ball from him and there it was: *10/21, property of Wyatt Nystrom.*"

"Cool," Wyatt said. He figured that it took six days for the ball to get from Eliza's house to Aunt June's school, a distance, he figured, of about three miles. He calculated that if it took six days for the ball to go three miles, theoretically, it could travel the entire sixty-mile length of the Brandywine in just a few months. *Great*, he thought. *The madness could be just a few months old, or from the eighteen hundreds. That really narrows it down.*

Check It Out!

"Also, my mom has the papers with all the data on it from her class's stream study. I'll bring it over when we come later. I can't wait for the Pumpkin Festival!"

"OK," said Wyatt. "We'll just collect the data until we can get to Dr. Flo's house and put it on the map."

~ ~ ~ ~ ~

While the computer booted up, Wyatt poured a bowl of cereal. Despite his protests, Wyatt's parents had insisted on setting the computer up in the kitchen, where it was *in the open.* He placed the bowl on the kitchen table and swiveled his chair around, having learned the lesson the hard way not to eat over the computer. The time he dropped his bowl of cereal onto the keyboard, he was picking out pieces of dried cornflakes for weeks afterward.

"Tennis balls," he googled. He glanced at the mystery white USLTA ball and the fluorescent yellow USTA ball he had placed side by side. He scrolled through the list of sites when "normally yellow in color" caught his eye. He clicked on the link. Early tennis balls, he found, were made of leather stuffed with hair or sheep stomachs tied with rope. He looked at his soggy cereal flakes. *I guess I'm not that hungry anymore,* he thought. He continued his search and found that the game of tennis began in the late 1800s. *So at least we know that the madness started after that.* Balls were colored fluorescent yellow after 1972 to make them more visible on television. *So this ball probably came from before 1972.* The only colors approved presently are yellow and white, according to the USTA. *If USTA is United States Tennis Association,* he thought, *what is USLTA?*

The front door opened. "Can you give me a hand, Wyatt?" his mother called. "And you've got to take Benny for a walk before the rest of the family gets here."

Wyatt left the computer to help his mother carry in the groceries.

~ ~ ~ ~ ~

Danni and her parents arrived with pizza and the data sheets from June's fourth-grade class. She handed the sheets to Wyatt.

Date: October 27
Time: 10:30 am
Location: Ring Run under Chadds Ford Elementary School bridge
Method of Sampling: general area search
Weather Conditions: warm and sunny, about 55 degrees
Weather in Last 24 Hours: Rain
Biotic Index: 10
Points of Interest: Route 1 (busy main road); school property. Note: water muddy
Study completed by: Mrs. Nystrom's Chadds Ford Elementary School 4th Graders

Checked off on the form were net-spinner caddisfly, mayfly, stonefly, riffles beetles, water pennies, planaria, and dragonfly larvae.

Sticker It Blue!

Date: October 27
Time: 11:30 am
Location: Brandywine Creek at Chadds Ford, just upstream of Route 1 bridge
Method of Sampling: general area search
Weather: warm and sunny, about 60 degrees
Weather in Last 24 Hours: Rain
Biotic Index: 9.
Points of Interest: utility lines, Route 1, wetlands, town of Chadds Ford, Brandywine River Museum, Brandywine Conservancy all in vicinity; water very muddy
Study completed by: Mrs. Nystrom's Chadds Ford Elementary School 4th Graders

Sticker It Red!

Fewer types of macroinvertebrates were checked off on this form.

"Didn't Dr. Flo say that what we want is a Biotic Index over ten? It looks like Ring Run is good, but not the Brandywine at Chadds Ford," said Wyatt. "Probably because of Route 1."

Danni nodded. "Here," she said, handing him the tennis ball. "I can't believe they found this!"

"It was hung up behind some very slimy leaves," added June. "You should have seen the water; it was really moving after the last few days of rain. And the kids said it looked like chocolate milk!"

Wyatt pictured the chocolate syrup and water mixture he had poured into Dr. Flo's pollution experiment at her house a few months back. He decided not to share the term "poopy water" with his aunt, remembering that the "chocolate milk" he and Dr. Flo had made was the result of various pollutants and untreated wastewater. After the last few days of rain, a lot of Aunt June's chocolate creek was probably from soil erosion.

"Thanks a lot, Aunt June, for doing this."

"Oh, it was a great field trip, practically in our own schoolyard. No buses, no cost; what's not to love? The kids were very intrigued by your letter and your request for them to help. It's *real-world* science. By the way, we'd all love to hear the results when you've figured out the *madness*," she said. "And I hope you don't mind that I forwarded the letter to a few other teachers who I'm sure would love to be involved."

"No, that's great. Thanks."

Wyatt's dad walked in the front door.

"Now, how about some pizza? I think we all want to get over to the Great Pumpkin Carve," said Meg.

~ ~ ~ ~ ~

The nighttime traffic at the intersection of Route 1 and 100 was completely backed up, so the Nystroms parked in a nearby restaurant lot. Guided only by oncoming headlights, they walked carefully along the narrow road. At the entrance to the fair, Danni and Wyatt took off as soon as their hands were stamped. "We'll see you later," they called to their parents.

Map It!

Danni and Wyatt found Eliza, along with two of her friends, waiting for them behind the funnel-cake tent. The full moon was low in the eastern sky and a few flickering jack-o-lanterns cast erratic light on their faces. Danni blurted out the news that her mother's class had found the tennis ball that Wyatt had launched from the stream behind Eliza's house.

"Slow down, Danni, geez," said Eliza.

"Sorry," Danni giggled. "But it's pretty cool. Do you realize what this means?"

"Wait a minute," Eliza said. She introduced her friends and told them about the tennis ball they had launched from the back of her house. Wyatt explained the significance of the traveling tennis ball in the context of the madness mystery, and recited the inscription dramatically, taking advantage of the eerie setting of the Halloween event.

"Creepy, isn't it?" Eliza looked at her friends, who nodded.

"Anyway," said Danni. "Now you can send messages to my mom to give to me. Just send a message down the creek behind your house," she said, laughing.

"What it really means," Wyatt added, "is that whatever happens in your neighborhood can affect the water that comes down to us in Wilmington, where it all seems to be polluted. If pollution is the madness, we want to find out where it's coming from."

"Don't look at me," said Eliza. "It's not like I put tennis balls and other garbage in the creek."

"Not just garbage." Wyatt explained some of the sources of pollution that originate on land. "Like salt spread on roads to melt ice in winter. Or fertilizer and weed killer people put on their lawns. Even washing your car in the driveway can cause pollution."

"Yeah," added Danni. "You don't have to be a factory to pollute. We're all doing it without even knowing it."

"But how does that stuff get in the creek from my backyard?"

Wyatt explained how rainwater washes pollutants from the land into the creek, and how the smallest of tributaries, like Ring Run, can then contribute those pollutants to the Brandywine.

"Wyatt, you should be a science teacher," Eliza said. "You can really explain this stuff."

Wyatt felt his face flush. Grateful for the dark setting, he changed the subject. "We've had enough school this week. Let's go check out the pumpkins."

The small group raced to the display of elaborately-carved pumpkins. Using their ballots, they cast votes for their favorites. Danni chose a four-hundred-pound pumpkin carved with monsters from *Where the Wild Things Are*, and Wyatt voted for the one with a scene of the headless horseman riding by moonlight. Before Eliza and her friends could decide, Danni heard her parents call.

"Guess we have to go," she said. "I'll see you tomorrow at art class, Eliza."

"Ditto," said Eliza.

~ ~ ~ ~ ~

"Let's cut through the field this time," suggested Wyatt's mother. "I don't like walking on the road. It's too dangerous."

The adults picked their way through the rutted field. Wyatt and Danni took off, slowing down when they hit a muddy patch. Wyatt took a step forward and realized his back foot was suddenly shoeless. His shoe had stayed behind, stuck in the mud. He reached back for it, pulled until the suction released the shoe, and, thrown off balance, fell into the mud. Danni laughed.

He called back to his mother. "Mom, I fell. We're going to the creek to wash off." Knowing his mother would be hesitant about letting him go to the creek at night, he didn't wait for her response. With a shoe on one foot and a soggy sock on the other, he dashed across the boardwalk to the creek. Danni followed behind. The creek shimmered with the reflection of the full moon. Wyatt peeled off his sock and rinsed it in the water.

"*Mushhakot.*"

Wyatt and Danni turned toward the strange sound and saw two dogs and several pigs sniffing the ground around a horse. On top of the horse sat an elderly woman with a long, gray braid. The moonlight cast an eerie bluish glow on the group.

"What?" Wyatt asked, not sure if the strange sound had come from the woman.

"*Mushhakot*," the woman said.

"*Mushhakot*?" Wyatt repeated.

The woman nodded, smiling. "The sky is clear."

"Yeah, I guess," Wyatt said.

Danni nudged Wyatt. "Let's get out of here," she whispered.

"Wait," said Wyatt, more curious than scared.

"Um, we gotta go. You know, our parents are waiting, you know, expecting us," said Danni.

"*Lapich Knewel*."

Check It Out!

Danni and Wyatt turned and raced back across the field, over the boardwalk and to the car. *There seems to be madness everywhere*, Wyatt thought. But neither of them uttered a word.

Field Trip!

November

Chapter 12

MARCHING ORDERS

Danni and the girls in her class dumped out a bucket of blue plastic toy soldiers. They were ready to play out the largest battle of the American Revolutionary War: The Battle of the Brandywine. Mrs. Bibbo gave the boys a bucket with an even larger number of red toy soldiers. She had told the class that each toy soldier represented about one hundred real soldiers in the actual battle.

"Girls are the American Patriots, boys are the British Crown Forces. Now, remember girls, you are trying to protect Philadelphia, where your new government is operating, from the British. Boys, you want to get past the Patriots and take over Philadelphia to put a stop to this whole revolution."

She rolled a large map out onto the classroom floor and directed the two groups to leave their soldiers in piles: the Patriots on the eastern bank of the Brandywine near Chadds Ford, and the British near Kennett Square.

Map It!

"Before we actually act out the conflict this morning and then go to Brandywine Battlefield Park this afternoon, let's review," Mrs. Bibbo said as she went to the blackboard. She wrote in large, neat cursive, *Battle of the Brandywine.* Then she added: *A Time of MADNESS.* Danni's eyes widened at the phrasing. *Maybe this is the madness,* she thought. *Maybe George Washington dropped the tennis ball into the Brandywine. Did George play tennis?* She looked out the window, thinking about this new possibility. There, standing in the parking lot, was an African-American man dressed in an old-fashioned soldier's uniform. He had a rifle slung across his shoulder and was eyeing the school. *Not again,* Danni thought. *Not another ghost!*

"Danni, pay attention," Mrs. Bibbo said. "All right, class, what led to the Battle of the Brandywine during America's fight for independence?"

Several hands shot up. Answers about the Declaration of Independence, the Boston Tea Party, Paul Revere, George Washington and the First Continental Congress were called out. Mrs. Bibbo wrote the students' answers on the board.

"Let's back it up just a bit, folks," Mrs. Bibbo said. "Before all that, William Penn was deeded the area we now call Pennsylvania in 1681 to establish a colony for Quakers to practice their religion freely. Do you remember what Pennsylvania means?"

"Penn's woods," said one of the students.

"Good," said Mrs. Bibbo. "Prior to and during this time, mills were built along the Brandywine and other rivers throughout the Northeast. Millers were becoming quite successful. And where there is successful business, there will always be . . ."

"Taxes," called out one of the students.

"Exactly. Now, let's put this in some kind of order." Referring to a handheld paper, she added dates to the events on the blackboard. That was one of the things Danni loved about Mrs. Bibbo; she let the class use references so they didn't have to memorize dates. "What's important to understand is the sequence of what happened in history and why those things happened," she had told the class earlier in the year. "Dates can be looked up. That's what references are for."

The class created a timeline starting with the colonization of America and taxation without representation, leading up to the Battle of the Brandywine and concluding with the ratification of the United States Constitution.

Check It Out!

"All right, class, let's get this battle started," said Mrs. Bibbo. "Danni and Johnny, I mean, General Washington and General Howe," she said and stuck a nametag on each of them. "Get your troops in order."

As leader of the Patriots, Danni commanded her troops to set up their soldiers. They placed most of them along the Brandywine Creek at Chadds Ford to block the direct advance of the British troops from Kennett Square to Philadelphia. To keep the British from crossing upstream, they placed a dozen more at Brinton's Ford and a few more just below the forks of the East and West Branches of the Brandywine.

Map It!

"Now, General Howe," Mrs. Bibbo said, as she huddled with Johnny and the Crown Forces. "You've got your troops of British, Hessian, Scottish and Loyalists gathered at Kennett Square."

"What are Hessians?" one of the boys asked.

"Hessians were German soldiers hired by the British to fight against the Americans. They were the ones who wore the argyle socks and the pointy fur hats to battle."

"That sounds more like a Halloween costume than a uniform," the boy snickered.

"Well, the Crown Forces were definitely more well-dressed than the Americans. That's because they had better resources. Anyway, your troops are recovering after a difficult journey by ship from New York to the Chesapeake Bay. Many of your horses died, many of your soldiers got sick and much of your food was spoiled. The Brandywine Valley has farms with livestock and wagons, mills for clothing, and food, so it's a good place to regain some of your lost resources. So you gather in Kennett Square."

Map It!

"How do you regain?" asked another of the boys.

"Well," Mrs. Bibbo hesitated. "You *steal*. You're going to steal horses, wagons, and food from the locals, and then you're going to destroy the mills so the Patriots can't get what they need."

"That's not fair," said Johnny.

"No, but it's war. And even though there were rules, many soldiers broke the rules. All's fair in love and war, you know.'"

"Yuck," said several of the boys.

"Let's finish up with your pep talk, OK? Now you want to get your troops from Kennett Square to Philadelphia to capture the center of American political control. But you have a major obstacle."

"The American army?" asked Johnny.

"Well, yes, the Continental army was a major obstacle. But the natural obstacle is the Brandywine Creek. From earlier battles, you learned to avoid a direct head-on fight, so you want to surprise the Americans by sneaking up behind them. The Americans are guarding all the fords where they think you can cross the Brandywine with your wagons. But you know that there are unguarded fords above the forks where you can cross."

Johnny divided the red toy soldiers equally into two groups, one led by a student designated as Lord Cornwallis and another as Lieutenant General Knyphausen.

"All right, troops, rise and shine," declared Mrs. Bibbo. "It's September 11, 1777. It's foggy and muggy and it's going to be a long, hot day!" She pressed the play button on her computer, and the Rocky theme song blared with a crescendo of horns, trumpets and tambourines.

Check It Out!

Johnny nodded to Lord Cornwallis who, with the help of several classmates, began moving their large group of red toy soldiers north. Knyphausen headed east with his soldiers, directly toward the bulk of Washington's troops at Chadds Ford. Danni nodded and the British and American troops began the fight west of the Creek across from Chadds Ford. Cornwallis, finding almost no American forces defending Trimble's and Jefferis' Fords, crossed his troops over the West and East Branches of the Brandywine.

Map It!

"Now, General Washington, you've been informed that the Crown Forces are approaching from the hills just north of Birmingham Friends Meeting House which you've been using as a hospital for your injured soldiers. If you don't change something, they will attack you from the rear."

Map It!

Danni immediately ordered about four dozen soldiers to pull out of the fight at Chadds Ford and get over to the Birmingham Meeting House, where the British, Hessian, and Scottish soldiers outnumbered them nearly two to one.

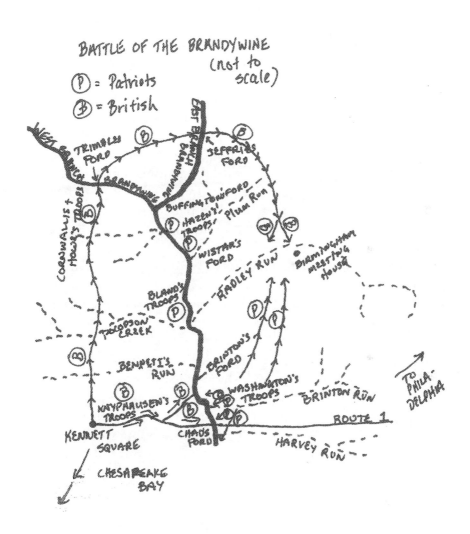

BATTLE OF THE BRANDYWINE

Try It!

"Now, everyone, sit, close your eyes and listen." Mrs. Bibbo hit play on her computer and the room resounded with the cracks and thunder of artillery and musket fire. Above the din, she shouted. "Imagine the battlefield. It is hot and muggy and the layers of your stiff, heavy uniform are weighing you down. Children your age are playing the fife and drum, indicating your position. Some of your enemies are no farther than fifty yards away, shooting directly at you with rifles, muskets, and cannons. The battle is raging, bayonets glinting in the sunlight, cannons aimed at your brigade, muskets firing all around. The deafening sound of gunfire surrounds you. Minutes turn into hours of battle. The air is acrid, thick and white, smoke-filled. You can barely see and barely breathe. It is four o'clock, five o'clock, six o'clock, and the battle wages on. Men drop by your side, blood oozing from legs, arms and skulls. Retreat is ordered. Cannons are abandoned, weapons discarded, soldiers are fleeing."

The classroom door flew open, slamming into the wall. Danni opened her eyes. An African-American man with glasses, a rifle in hand, and dressed in breeches, a waistcoat and a cocked hat, shouted. "Everybody, down!" Danni giggled, relieved that the man she had first spotted in the parking lot was not a ghost after all.

"Are you Patriots or Loyalists?" he asked the students.

"Go ahead, answer him," said Mrs. Bibbo.

"Patriots," Danni and the girls shouted back.

"Ah, good," he replied. "Then let's get to work." He stopped suddenly and looked at the map on the floor. "What is this, pray tell?"

Danni raised her hand. "We were acting out the Battle of the Brandywine."

"Really?" he said, confused. "Acting it out?" He glanced at her nametag. "So glad to make your acquaintance, General Washington," he said. He pulled an official paper from his vest pocket and handed it to her. Turning to the rest of the Patriots, he said, "Now, excuse me while I make haste to gather the discarded weapons of our fallen brothers before they fall into the hands of the enemy." He scrambled around, miming the gathering of muskets and rifles. Let's go," he yelled, dropping the imaginary weapons into a cart and running toward the door.

And then, the man was gone. The class sat silently for a few moments.

"Who was that?" one of the girls asked.

Mrs. Bibbo nodded to Danni, who looked at the paper in her hand. She read slowly. "September, 1777. Let it be known that Edward 'Ned' Hector has been assigned to the Third Pennsylvania Artillery Company of Colonel Proctor's Regiment in Brandywine, Pennsylvania, for the defense of Philadelphia, where he is to serve to halt and repel the advancing British forces. Signed, Continental Congress, George Washington, Commander in Chief of the Continental Army." Danni looked up. "Hey, that's me," she beamed.

A knock sounded at the door. Mrs. Bibbo opened it to find Ned Hector.

"May we reacquaint ourselves, madam?" he said. "Our first meeting was so rushed by dire circumstances."

Mrs. Bibbo led him in. "Class, will you please welcome Mr. Ned Hector." The students burst into applause and the actor bowed deeply.

Check It Out!

"Ah, thank you, dear children," he began. "You are most generous to welcome me back after such a frenzied first acquaintance." Mr. Hector sat in a chair. "I see you are studying a most important chapter in our American history. Pray tell, can you inform me as to the outcome of this war?"

"We won," the girls called out together. Some of the boys scoffed.

"Ah, yes we did," he agreed. He questioned the students on their knowledge of the Revolutionary War and America's fight for independence. He explained his uniform, including his hat, cocked to avoid interference with the rifle he had resting on his shoulder. He contrasted elements of modern life with the late 1700s: cars replacing horses and wagons; girls schooled alongside boys; daily hygiene evolving from bathing only twice a year. He shared some lesser-known facts about the role of African Americans and women as soldiers, wagon drivers, nurses, and launderers during the war. "We were a ragged lot, our American forces. Up against the world's best trained army, we were colored and white, free men and slaves, farmers and merchants, all fighting for the common cause of freedom from an English king. *E pluribus unum*," he added. "The many that become one."

Ned Hector bowed again graciously. "Good day, my fellow Americans. It is a good day. Or, as my Lenape friends would say, *Weli Kishku.*"

Danni shivered at the bewildering words. An image of the woman on horseback came to mind.

Check It Out!

~ ~ ~ ~ ~

The girls from Mrs. Bibbo's class hiked up the hill and then ran down the other side as they headed toward the final station of their visit to the Brandywine Battlefield. After learning about colonial dancing, Revolutionary War weaponry, colonial medical treatments, iron forging and everyday life of the simple and peaceful Quakers who refused to participate in the war, but whose homes quartered important members of the Patriot cause, the girls entered the Visitor's Center museum and, of course, the gift shop.

Map It!

A map and timeline of the events of the Brandywine Battle adorned the wall of the museum, ending with the British occupation of Philadelphia fifteen days after the September 11th battle.

"What happened to the capital if the British captured Philadelphia?" Danni asked.

The guide turned to her. "Well, they relocated to the safety of Lancaster for one day," she answered. "Then they moved out to York."

Danni wandered around the museum. A large framed poster of soldiers entering a fight caught her attention. A motley crew of young, old, portly, thin, decorated and ragged men were armed with fife, drum, bayonet, and rifle. They were real men, not toy soldiers, marching directly into battle, driven by determination and courage to create a new nation. Danni was surprised by her emotional response to the picture. *Now that's a powerful image*, she thought, quoting her art teacher. She read the label beside the poster. THE NATION MAKERS/

HOWARD PYLE, 1903/ARTIST OF CHADDS FORD AREA/WELL KNOWN FOR HIS HISTORICAL ILLUSTRATIONS AND CREATION OF DRAMA/ TEACHER OF N.C. WYETH.

The war, the Brandywine, the artists, Danni thought. *They're all connected. Jeesh, I'm beginning to sound like Dr. Flo myself!*

In the gift shop, Danni spotted Mrs. Bibbo talking to a woman she didn't recognize.

. "Danni, come meet my former student teacher," Mrs. Bibbo called. "Miss Speck teaches fifth grade at Starkweather Elementary and is very interested in your mystery on the Brandywine. I was telling her that our class is trying to get down to the creek soon before it gets too cold to collect data for you. She's interested in doing it with her class, too."

"Yes," said Miss Speck. "I've actually done some research myself to help clean up some of the polluted parts of the Brandywine," she explained. "The Red Streams Blue project was created to change the red, or polluted, parts of the stream to blue, which means clean." Danni thought it interesting that *red* meant bad and *blue* meant good, like their British and American toy soldiers. Miss Speck continued. "I'd love to have my students help with your research, if you're still looking for help, that is."

Check It Out!

"Oh, definitely," Danni responded. "I'll tell my cousin Wyatt, and we'll send you the stuff that tells you what to do." *I'll have my people get in touch with your people,* she thought to herself.

Field Trip!

Chapter 13

QUEST

The Nystrom's minivan climbed the hill toward Pocopson Elementary. As soon as Luke parked the car, Danni and Wyatt scooted out to drop off a water-quality packet for Mr. Campbell. A teaching friend of June's, Mr. C had promised to collect data with his class to help with the mystery. The cousins could see why this was convenient. Pocopson Elementary overlooked the Brandywine Creek. Except for the road at the bottom of the hill, and the cars bustling in and out of the hardware store parking lot, it would be an easy walk for the students to get to the creek.

Map It!

They turned out of the school's drive. The clanging bells at the railroad crossing quieted and the barriers lifted. A long freight train, with its cars and flatbeds, rumbled alongside the road, making its way from Wilmington to Coatesville.

"Hey, Lenape Road," Wyatt said as his father turned. "Wasn't that Indian lady Lenape? You know, the one that had that statue near Longwood?"

"Yes, the Lenape lived all over this area," Aunt June answered. "Unfortunately, they were pretty much driven out by the colonists. Don't you learn about any of this in school?" she asked. "I teach it in Pennsylvania."

"Maybe we did," answered Wyatt. "I probably just don't remember."

"Yeah, you were probably daydreaming about kissing Jennie," Danni teased.

Wyatt punched her playfully on the arm. "*Ixnay*," he added, not realizing that his parents had used pig Latin in their childhoods.

"Well, they should teach it in a way that you would remember," Meg said. Wyatt felt relieved that his mother didn't seem to notice Danni's remark about his love life. Hadn't he told her to keep it quiet?

"The Lenape Indians were very important," she continued. "Where do you think Pocopson School got its name?"

"From Pocopson Township," answered her husband. "That's kind of obvious."

"Well, then, where did the township get its name?" she said, rolling her eyes. "It's a Lenape word."

"What does it mean?" asked Wyatt.

"Roaring waters," Meg answered. She turned and winked to the kids in the back seat, pointing at the Pocopson Nursing Home brochure she held with its "Interesting Facts."

They turned into the nursing home driveway, ready to check out the possible new residence for Grandma Nystrom.

"They don't look like any *roaring waters* to me," said Uncle John as they crossed over a small creek.

Map It!

~ ~ ~ ~ ~

After spending most of the tour indoors, Wyatt was eager to get on with the day and spend some time outside. They left the nursing home and turned onto the main road. "*Wawaset*," Luke said, noting the name of the road. "That sounds Indian. What does that one mean, Miss Smarty Wife?"

"I think it means *land of hunger*," June said, laughing, before Meg had a chance to answer. "I'm ready for some lunch."

The car puttered along the backcountry roads, passing vast open spaces of land, peaceful and quiet except for the intrusive gates of Chester County Prison. After a short drive, they reached Northbrook Market Place for lunch and then headed over to the Myrick Center for a hike.

Map It!

~ ~ ~ ~ ~

The sign at the entrance to the gravel drive read BRANDYWINE VALLEY ASSOCIATION/RED CLAY VALLEY ASSOCIATION/MYRICK CONSERVATION CENTER.

Map It!

"That's a mouthful," remarked John Nystrom. "I wouldn't want to have to be their receptionist."

"Why is their name so long?" Wyatt asked, as he got out of the car.

"Well," a voice from behind him said.

"Dr. Flo," Wyatt shouted. "Danni, Dr. Flo's here with Benny!" The dog ran to Wyatt, wagging his tail.

"Your mother asked if I could check on Benny, since you'd be gone all day. When she told me you were heading here for a hike, I decided I'd join you. I didn't feel the need to go to the nursing home," Dr. Flo chuckled, "*yet*. Anyway, I figured that Benny would love to come out to the Myrick Center and run around."

"Cool," said Wyatt. The air had crept up to seventy degrees, unseasonably warm for November. Benny panted.

"Now, he's got to be on the leash until we get far enough from the road. We don't want Benny or any of you to become road kill."

"So, why is the name of this place so long?" Wyatt asked Dr. Flo, realizing she had never answered the question.

"Well," Dr. Flo said as they crossed the road. "This used to be Horatio Myrick's farm. When he died, he left the property to the Brandywine Valley Association, to keep it conserved. The Red Clay Valley is the 'next-door' watershed. It's quite a bit smaller than the Brandywine Valley, so they share BVA's staff and programs."

"This is a heck of a piece of property to leave to an organization," Luke said. "He could have sold it to a developer and left a lot of money to his kids."

"He didn't have any family," said Dr. Flo. "He was an eccentric man who lived alone with his cows. And I don't know if the cows would have put an inheritance to good use."

"What's a watershed?" asked John. "You keep mentioning that."

"Wyatt, will you do the honors?" Dr. Flo looked to Wyatt, who explained to his uncle that a watershed is an area of land that drains into a particular body of water.

"In this area, water drains into the Brandywine Creek," Wyatt said. "So anything you do to the land in the watershed will affect the water in the creek."

"The Brandywine Watershed encompasses over three hundred square miles that drain into sixty miles of the Brandywine Creek, from Honey Brook to Wilmington," Dr. Flo added. "Since we all live in Wilmington and get our water from the Brandywine, we're basically the recipients of all that's done to the water, and the land, upstream. Wyatt is right on target when he says that to protect the water, you must protect the land around it. That hasn't always been done."

"I know you're a doctor and all," said Wyatt, "but how do you know so much about this place and about watersheds?"

Dr. Flo looked at Wyatt and smiled. "Do you remember the portrait of my Uncle Clayt in the living room? His full name was Clayton Hoff, and he was one of the founders of the BVA, along with Robert Struble, Sr., *way back in forty-five*—1945, that is. Uncle Clayt used to always say it that way, 'We got BVA started *way back in forty-five*.'"

Forty-five, forty-five, why does that number ring a bell? Wyatt thought. An image of the woodblock and the rubbings popped into his mind. *No, that was forty-four. But it sure is close.*

"Cool," Danni said, poking her head around a door into the Browning Barn. It was obvious to Wyatt that Danni was impressed with the barn, not the information about the founding of the BVA. He followed her inside to a massive space lit only by beams of sunlight streaming in through cracks in the walls.

"It's warmer outside," Wyatt said. They wandered around the rooms of the chilled barn and passed a faded picture of a hunched, elflike man standing beside a cow.

"He looks weird," Danni said.

Between two doors marked BUCKS and DOES hung a large painting of a curly-haired man with round-framed glasses: TED BROWNING, 1938-1987, CONSERVATIONIST, OUTDOORSMAN, NATURALIST, FRIEND TO BVA.

Check It Out!

"He looks nice," Danni said, continuing her evaluation of BVA characters.

Dr. Flo hesitated. "He was a wonderful man and gifted at so many things. His death was a terrible loss to the community."

"Is the barn named for him? It's called the Browning Barn," asked Wyatt.

"Yes, it is. A perfect tribute to a man who loved this place."

"Yuck," said Danni. She had peeked behind the door marked "Buck" and spotted the no-flush toilet. "It's an outhouse, only in the house."

"*Clivus Multrum*," said Dr. Flo.

Wyatt stiffened, thinking she sounded an awful lot like the woman with the long gray braid who spoke with a foreign tongue. "Did you just speak in another language?" he asked.

"No, silly. A *Clivus Multrum* is a composting toilet. It's environmentally friendly. Waste drops into a bin under the barn, where it composts over time. Eventually, it can be applied to the land for its nutrients. Human manure."

Check It Out!

"Humanure," joked Wyatt.

"That's just gross," said Danni.

"It's just different, Danni," said Dr. Flo. From high above, shrill twittering sounds trickled from the rafters as tiny pellets dropped to the floor like raindrops. "Like that bat guano. You may think it's vulgar, but, in many cultures, bat and bird guano is used as a natural fertilizer for crops."

"What's guano?" Danni asked. "Although I'm not sure I really want to know."

"Feces, excrement, solid waste," Dr. Flo paused, searching for appropriate words that Danni would understand.

"You know, Danni," said Wyatt.

"It starts with an 'S'
Ends with a 'T;'
Comes outta you
And it comes outta me."

Rap It!

"Wyatt!" Dr. Flo reprimanded.

"Oh, *scat*," Danni remembered from her art lesson.

"That's not what I thought you were going to say," Dr. Flo said, winking. "As for our *scat*, if that's what you'd like to call human waste, it is treated rather inefficiently in our culture."

"Don't tell my dad that," said Wyatt.

"Yes, I know he manages a wastewater treatment plant and it's respectable work. I don't know where any of us would be without the wastewater workers of the world. But it is an inefficient way to treat human waste. Wyatt, you must know a little about how a wastewater treatment plant works."

"Well, basically, you flush the toilet and water goes from your house, through underground pipes, to the treatment plant, where the waste is taken out and sent to the landfill. The water goes back into the creek."

Check It Out!

"Yes, exactly," said Dr. Flo. "Water carries the waste from our homes and schools to a treatment plant. There, the waste is filtered out, but it still needs to be disposed of. And the water that carried the waste must be cleaned before it can be put back in the creek."

"So the water just carries the waste?" Danni said.

"Yes. So it's a lot of work, just to carry waste from your home. The Clivus Multrum, on the other hand, is a system that doesn't need water at all. And the result is a useful product: fertilizer."

"Yeah, but what about the stink?"

"Do you smell anything?" said Dr. Flo.

~ ~ ~ ~ ~

"Oh, my goodness, that sun feels good." Outside the barn, Dr. Flo stood facing the sunshine. The four Nystrom adults stood waiting outside the barn as Danni and Wyatt joined them.

"Let's get going," Meg said. "Benny is dying to go exploring. What have you been doing?"

"Oh, just talking scat," Wyatt joked.

They followed a wide dirt path that led uphill. Danni peeled off her sweatshirt and tied it around her waist as they reached the top of the hill.

"Do you see why they call it *Skytop*?" Dr. Flo asked.

"What a view," said Meg. "It's lovely."

"Guess what we just passed over, Wyatt," said Dr. Flo.

"Um, a hill?"

"Yes, a hill, which is a highpoint, so must be a . . ."

"A watershed boundary," Wyatt said.

Spread before them, a golden field dipped toward a grove of barren trees and rose again on the other side. To the south, trees lined the top of a hill. From the north, a neighborhood of wealthy homes at the top of another hill overlooked the grove, so the entire area was like an immense bowl with the grove of trees at the bottom. Benny immediately took off down the hill. Chasing after him, Wyatt and Danni stumbled on the uneven ground, fell and rolled down the hill, laughing. Benny raced around the grove of trees and up the far hill when Wyatt caught a glimpse of what had attracted the dog.

"Fox," Wyatt called. The group watched Benny pursue the fox up the far hill and disappear out of sight.

Wyatt knew Benny would return eventually. He and Danni sprinted the rest of the way down the hill. By the time the adults arrived, the cousins had discovered a tiny creek.

Dr. Flo gestured to the surrounding landscape. "All this land, to the tops of the hills around us, drains into this little stream. If we call it the Wyatt Stream, then this area would be the Wyatt Watershed."

"That's cool," said Wyatt. Not in the mood for another lesson, he scampered down a narrow trail. Danni was right behind him. "C'mon," he called to the others. The trail led to another creek that merged with the first one. The kids splashed water on their faces.

"It's so hot out," Danni said. "I can't believe it's November." She pulled a rock out of the small creek. "Look, Wyatt, what's this called again?"

Wyatt peered at the fingernail-sized invertebrate squiggling along the rock's bottom. "Three tails like the three letters in *May* is a mayfly," he answered with confidence. "And there are some caddisfly homes, too." He pointed out the small shelters attached to the rock. "All good guys."

~ ~ ~ ~ ~

Benny suddenly appeared and headed directly toward Wyatt and Danni.

"Where have you been, buddy?" Wyatt asked. He picked a sticky burr from Benny's fur. Looking closely at the seedpod, he noticed tiny hooks that had stuck to Benny's coat. They reminded him of Velcro. He picked at the burrs, but gave up when his mother put the leash on Benny to cross back over the road.

Check It Out!

~ ~ ~ ~ ~

Acorns crunched underfoot as the troupe passed under the huge oak tree by the stone springhouse.

"That oak tree was here when William Penn was granted this land in the late 1600s," Dr. Flo informed them. "It's about three hundred and fifty years old and still producing nuts like you."

"Very funny," said Wyatt.

Dr. Flo walked over to a mounted wooden box and pulled out a paper. She handed it to Wyatt. "I think your folks and I are worn out, but why don't you two try this. It's like a treasure hunt."

Wyatt looked at the paper titled "Myrick Center Quest." A crude map accompanied a poem filled with clues of where to explore.

> *"Welcome to our Quest*
> *The first in our state,*
> *At the former home of Mr. Myrick.*
> *His was a large estate.*
>
> *The Brandywine Valley is an illustrious place,*
> *Full of du Ponts and Wyeths.*
> *When the Hessians were here in 1777,*
> *They made sure that they passed by it."*

"Hey, the Hessians. I know them. They fought for the British during the Revolutionary War," said Danni. "Let me see that." Wyatt handed her the poem and she read on.

"Start by facing the stone springhouse with the 'Penn Oak' on your left,
They both have seen many years.
The Hessians took some pies and Mr. Myrick's ghost,
Has paid a visit here."

"What about pies and Hessians and ghosts?" asked Wyatt.

"We learned about that," said Danni. "When the British landed in Maryland so they could invade Philadelphia, they lost a lot of their horses and food, so they had to steal from the local people. I guess they stole pies right here, from that springhouse."

John Nystrom called the children over to a display board filled with information about the BVA. "It says here that the British soldiers wolfed down a week's worth of pies from the springhouse and then napped, giving General Washington's troops an opportunity to scout them out and avoid a battle."

Yeah, well, they should have scouted out more than just this spot, Danni thought, remembering the outcome of the Brandywine Battle. She pictured the scene: British and Hessian soldiers in their bulky uniforms, adorned with fur and polished buttons, snoozing in the shade of the mighty oak. Rifles and shotguns, resting across their bellies, rising and falling with the soldiers' steady breaths of deep sleep. As she imagined them lying side by side between the oak tree and the springhouse, she noticed a plaque on the ground. She brushed off a few leaves and acorns. "Horatio Myrick, 1899-1980."

"Is he buried here?" she asked.

"Who? Ray Myrick?" asked Dr. Flo. "No, but his ashes are scattered here. And as for his ghost, well, maybe you shouldn't have called him *weird*," she added. "You never know who's listening."

~ ~ ~ ~ ~

By the time Danni and Wyatt deciphered the clue of the Quest that sent them to the woods, they had hiked up a trail, down and around a pond, and back up a field. "Watershed boundary," Wyatt called each time he felt his pulse quicken as they reached a highpoint. Their upper lips and foreheads beaded up with sweat from climbing the hills in the increasing humidity. Overhead, turkey vultures circled and clouds of gnats hovered. In the distance, a chubby groundhog scampered across the field and disappeared in a hedgerow.

They entered the woods but the bare poplar and beech trees provided no relief from the growing humidity. Thunder grumbled in the distance.

"We should hurry up," said Wyatt.

"But I want to find the clay," Danni protested. "The Quest says it's near the wooden platform." She headed down a hill. Benny ran ahead and stopped at a small creek to lap up a drink. Danni crossed over the tiny trickle to examine the fallen tree described on the Quest paper. She dug out a small clump of clay from between the upended tree roots. "Look." She held up a ball of clay to show Wyatt.

A slight rustling in the leaves on the other side of the decomposing tree drew their attention. Danni hoisted herself up. Benny had already circled around the tree and sat obediently beside a slight, elderly man with a hunched back and a ruddy complexion. The man petted a calf lying by his side.

"Oh, hi," Danni said. Wyatt poked his head around the roots of the tree to see who Danni could possibly be befriending out in the middle of nowhere.

The man didn't respond. He just stared off into the distance. Danni tried again, more loudly. "Hi," she shouted. The man remained unresponsive, but continued to pet the calf.

"Maybe he's deaf," she whispered to Wyatt.

Wyatt stepped around the tree to face the man directly. "Are you all right?" he asked.

The old man turned toward Wyatt and nodded. The hair on the back of Wyatt's neck rose. He retreated back around the tree. Danni motioned for him to get out his cell phone and take a picture.

As he reached into his pocket, his phone rang with his mother's ringtone, the voice of Lily Tomlin. "One ringy dingy," said the nasally voice of the comedienne, completely incongruous with the gravity of the present situation. Wyatt snapped a shot of the old man and the cow, then answered the phone.

"Storm's coming," Meg said to her son. "Come on back now. It's looking a little ominous."

"Sure does," Wyatt answered. *In more ways than one.*

He and Danni turned to the old man to see if they could help him return to his car or his home or whatever, but he and the calf had vanished.

"Where'd he go?" Danni asked.

Wyatt shrugged his shoulders. Then he pulled his phone back out to view the picture he had taken.

"He was weird," Danni said, suddenly realizing that she had used that description very recently.

"That's funny," Wyatt said. Danni looked at the screen. There was the fallen tree and Benny. But where the old man and the calf had been, only a wispy white haze lingered.

Field Trip!

Chapter 14

GO WITH THE FLOW

"We think it was another vision," Wyatt confessed to Dr. Flo once he and Danni were alone with her in her study. He had told Jennie and Rob to explore the kooky hoose on their own for a few minutes so they could talk privately with Dr. Flo. The last thing Wyatt needed in middle school was to be known as a weirdo for seeing dead people. He pulled out his cell phone and showed Dr. Flo the ghostly image where the old man and calf had been.

"And you're sure he didn't just leave when you weren't looking?" Dr. Flo said. "Or that isn't a picture of lightning?"

Wyatt explained that they were in a hurry because of the approaching storm, but that there was no way they wouldn't have seen an old man and a calf heading out of the woods. And, no, lightning hadn't struck the spot. Or his head.

"The old man might be quicker than you'd expect," said Dr. Flo. "But the calf would definitely be hard to miss. Let's just check something."

Dr. Flo typed *www.brandywinewatershed.org* onto her laptop keyboard. Laughter drifted in from the direction of the bathroom.

Rob and Jennie probably just checked out the toilet, thought Wyatt. *I wonder if either of them dared to use the water in the toilet sink.* He looked at the computer screen where Dr. Flo had navigated to BVA's history link and a picture of an elfish man. "Is this the man you saw with the cow?" Dr. Flo asked Wyatt and Danni.

Their eyes widened and they both nodded.

"This is Horatio Myrick, better known as Ray, the man who donated the property to the BVA."

"And the man whose ashes are spread under the big oak tree," Danni said.

"That's right, Danni. You know, you two aren't the only ones claiming to have seen his ghost."

"This place is *sweet*!" Jennie said as she and Rob entered the study. Danni and Wyatt moved purposely to the living room, getting away from the computer screen with its image of Horatio Myrick.

Danni led the friends over to the wall map. She pointed to a sticker designating Dr. Flo's house in the center of Wilmington. "You are here." She proceeded to point out various landmarks so Jennie and Rob could become familiar with the layout of the Brandywine watershed. Starting at the confluence of the Brandywine with the Christina, she traced the creek upstream to Brandywine Park and Warner Elementary, Alapocas Woods and Wyatt's home, A. I. du Pont Middle School, Hagley Museum, Winterthur, Brandywine Creek State Park, the Delaware/Pennsylvania State Line, Chadds Ford, Longwood Gardens, Brandywine Battlefield, and the confluence of the East and West Branches of the Brandywine. "Then you can go up the East Branch of the Brandywine," she continued, pointing to various locations. "To West Chester, Downingtown, and, finally, Honey Brook. Or you can follow the West Branch to Embreeville, Coatesville, and Honey Brook." She stopped suddenly. "Oh my gosh, Wyatt!"

"What?"

"The Brandywine. Look at its shape."

Wyatt looked a little confused.

"It's kind of shaped like a *Y*," Jennie said.

"Oh, wow," Wyatt said. "Like on the woodblock."

Map It!

The group stood in silence as they all absorbed the idea that the *Y* might represent the entire Brandywine. Or maybe it was, as Wyatt had once suggested, an abbreviation for *Why*? Or maybe it was both.

"How come they both end up in Honey Brook?" Jennie asked, breaking the silence.

"They don't end up in Honey Brook," Dr. Flo explained. "That's where the Brandywine Creek begins. It ends up in the Christina River in Wilmington."

"But why does it start out as two creeks from the same place?" asked Rob.

"Um," Wyatt murmured, looking at Dr. Flo for help.

"Now, Wyatt, you should know that," said Dr. Flo. "What must be between the headwaters of the East Branch and the West Branch?"

Danni brightened. "Highpoints," she blurted, pleased to beat Wyatt to a watershed answer.

"Absolutely. The creek that you see down in Wilmington is the result of the accumulation of hundreds of tiny headwaters throughout the Brandywine Valley. The headwaters of the two main branches, the East Branch and the West Branch, are located in many places; Honey Brook is one place where there are headwaters of both branches. The reason they go in different directions and separate from each other is because there must be highpoints between them." Dr. Flo paused, then added, "Wyatt, lie down."

Wyatt did as he was asked and lay down on the rug.

"Now, put your hands over your head and raise your arms slightly, with your fingertips spread out and higher than your wrists. Your hands should almost, but not quite, touch." She let Wyatt adjust himself. "Behold the Brandywine," she announced.

"Oh, I see." Danni described Wyatt's makeshift watershed. "The fingertips are the headwaters. One arm is the East Branch and one arm is the West Branch. Wyatt's chest is where the East and West Branches come together to make the main part of the Brandywine."

Rob joined in. "So the fingertips are in hills, right? Because they're up higher. And there must be highpoints between all the fingers."

Try It!

"By George, I think you've got it," said Dr. Flo. From the coffee table drawer she pulled out turquoise and royal-blue markers. "Now, try this. Danni and Rob, go get a couple of chairs from the other room to stand on. Mark every little stream that feeds into the West Branch with this royal-blue marker. Wyatt and Jennie, you use the turquoise marker to do the same on the East Branch. That should keep you all busy for a while! I'll go whip up a batch of my semi-famous oatmeal chocolate chip cookies."

Map It!

~ ~ ~ ~ ~

"All right, let's see what we've got here." Dr. Flo placed the cookies and a small stack of papers on the coffee table. She stood back from the map with the kids.

"Very nicely done. Now that you can tell which is the East Branch and which is the West Branch, find the highpoints between the two branches in Honey Brook, and mark them with an 'X.' Wyatt, do you remember how to read the contour lines and find the highpoints? Remember *knuckle mountain?*"

Wyatt nodded and showed the others how to locate highpoints by finding the little brown circles between the two branches of the creek. He pointed one out that was designated with a *BM,* next to the number *760.*

"I hope that doesn't mean what I think it means," Rob said.

"A BM is a *benchmark,*" said Dr. Flo.

"Yeah, it's not *scat,*" joked Wyatt.

"It's a location where they actually surveyed the height," Dr. Flo said. "A lot of times, benchmarks are at highpoints. Wyatt, do you remember any of the benchmarks around Winterthur?"

Wyatt stepped off of the chair to look at the area toward the bottom of the map, far south of Honey Brook. "Three sixty. That's way lower than up in Honey Brook."

"Which tells us which way the Brandywine is flowing."

Jennie, Rob and Danni all looked perplexed. Wyatt explained. "If it's seven-sixty in Honey Brook and three-sixty in Winterthur . . ."

"Seven-sixty what?" Danni asked. "Three-sixty what?"

"Oh, sorry," Wyatt said. "Feet. If the elevation is seven hundred sixty feet in Honey Brook and three hundred sixty feet at Winterthur, then the Brandywine must flow from Honey Brook to Winterthur, from the higher place to the lower place."

Try It!

"Which must mean that the creek starts there," Jennie said, pointing to Honey Brook, "and ends down there." She pointed to Wilmington. Wyatt nodded, appreciative that Jennie actually seemed interested in this whole thing instead of just thinking he was a map nerd. He had been a little nervous inviting her to Dr. Flo's, but now he thought it was turning out to be a good idea.

"Yes, and it is those highpoints that determine the fate of the water quality downstream," said Dr. Flo.

"Because what you do to the land, you do to the water," Danni and Wyatt said simultaneously.

"Now, let's take a look at what your Creek Pals have sent to us. These are the results that were sent to me."

Wyatt paged through the papers. Dr. Flo pulled a set of blue and red stickers out from the drawer that Wyatt was beginning to think of as an office supply store. Whatever they needed always seemed to be in that drawer.

"For any test result with a Biotic Index of less than ten, use a red sticker." Dr. Flo pointed to the sticker placed at Brandywine Creek State Park, where Wyatt and Danni had done their first official water quality assessment. "For any result with a Biotic Index greater than ten, use a blue sticker. That way, we can tell, at a glance, which parts of the Brandywine are in good condition and which aren't."

Jennie, Rob, Danni, and Wyatt all reviewed the various data sheets. They recorded the results and placed the stickers on the map at the appropriate sites.

Red Stickers	Blue Stickers
Christina River; BI= Polluted	Wilson Run; BI=14
Brandywine Park; BI=5	Ring Run; BI=10
Brandywine at Hagley Museum; BI=6	
Brandywine Creek State Park; BI=8	
PA/DE State Line; BI=9	
Brandywine at Chadds Ford; BI=9	
Brandywine at Pocopson; BI=6	
Radley Run near West Chester; BI=4	
Plum Run near West Chester; BI=5	

Sticker It Red or Blue!

When they completed the task, they stood back and looked at the stickers. It was clear that there were lots more red stickers than blue ones.

"So, what do we know about the mystery?" asked Rob.

"If the *Y* in the poem means the whole Brandywine," said Danni, "we've only got half of the information we need. We have to test on the East and West Branches, too."

Wyatt was upbeat. "Yeah, but look at what we do know. All of the test sites below the confluence of the East and West Branches of the Brandywine are red, except for some of the tributaries. The main stem of the Brandywine isn't as clean as it should be."

"What's a confluence?" asked Rob.

"It's where tributaries come together to make a bigger stream. In this case, it's where the East and West Branches of the Brandywine meet."

"What are tributaries?"

"They're smaller streams that come together to make larger streams," Wyatt answered. "Look." He pointed to each red sticker with a Biotic Index of *8* or *9*. "Even though they're red, these parts are not as bad as other parts. And they're either going through a park or just below wetlands." He pointed to the stickers marked *Brandywine Creek State Park, PA/DE State Line* and *Brandywine at Chadds Ford*.

"Remember the wetlands where we did our drawings for art class?" Danni said. "That's where we saw the bald eagles and Tina said that was a sign of clean water."

Jennie joined in. "And the cleanest tributary is Wilson Run that goes through Winterthur, where there's a lot of green on the map. Green must mean something good."

"Green means recycling the dying stuff and bringing it back to life," said Danni. "Like witches do."

"Now you're cookin'," Dr. Flo said, clapping her hands. "You are well on your way to figuring *something* out!"

~ ~ ~ ~ ~

"He looks nice," Jennie commented, passing the portrait of Uncle Clayt on her way to the front door.

"I adored my Uncle Clayt," Dr. Flo said.

Wyatt looked at the portrait, noticing something he hadn't before. One eye was closed just slightly more than the other. *Is he winking?* Wyatt thought, but chose to keep the notion to himself.

PART II

ABOVE THE CONFLUENCE

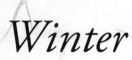

Winter

December

Chapter 15

ROCK LEGEND

A flurry of snowflakes splattered against the windshield as Luke Nystrom pulled into the lot of the nature preserve. Wyatt rolled down his window so Benny could stick his snout out and bite at the snowflakes. Just as quickly as it had arrived, the snow shower ended, and the early December afternoon turned sunny. Luke parked the car and Wyatt got out to stretch before hiking with the family.

"C'mon, boy," Wyatt called to Benny, releasing him from his stay command. The dog jumped from the car and ran to Wyatt. Eager to let his dog explore the open meadows and rolling hills of the Stroud Preserve, Wyatt broke into a jog.

Map It!

Luke and Meg Nystrom strolled behind as Wyatt and Benny crossed the stone bridge over the East Branch of the Brandywine. Along the dirt road, Wyatt observed hundreds of tree seedlings, each protected by a slender plastic tube. From the top of each tube, the seedling emerged with its baby branches and tiny leaves. *They look like birthday candles*, Wyatt thought, *for a really old person*. Benny stopped to sniff the air and then bolted across the open field. In the distance, spots of white flashed in the dull winter landscape. Wyatt recognized the telltale sign of the white-tailed deer signaling danger. He watched as they bounded silently toward the woods, Benny chasing. He knew that his dog would return eventually, so he made his way across the field.

At the creek, Wyatt shed his socks and shoes, rolled up his pant legs and stepped into the gurgling water to examine rocks. By the time his parents caught up to him, he had found caddisflies, stoneflies, mayflies, a water penny, several clams, a scud, and a leech.

"What are you doing?" exclaimed Meg Nystrom to her son, who stood barefoot in the stream. She had just left tracks in the light coating of snow.

"It's OK, mom, I'm not cold, really. And I'm finding all sorts of good macros here." He lifted another rock. More caddisflies and a snail. *Ten points, at least*, he thought. *OK, now I'm getting cold.* He hopped back to shore and sat on a rock, bending over to blow warm air onto his shriveled toes.

"You can't punish a kid for testing water in December," Luke said. "Even if he is a little obsessed."

"Is this the Brandywine?" Meg asked. "I saw you studying the map in the car."

"This is the East Branch," Wyatt said. "Above the confluence with the West Branch. All the data we've collected so far are from below the confluence. Dr. Flo says we need more data from upstream, so I'm collecting some. I'd give this site a Biotic Index of ten, though I didn't really do a standard test."

Meg and Luke Nystrom looked at each other. For a kid who didn't exactly thrive in school, it was obvious that he knew what he was talking about when it came to water quality.

Benny suddenly appeared and charged toward them. He dashed into the creek, chasing and biting at the splashing water, and then trotted over to his family. Shaking heartily, he sprayed water in all directions.

"Benny!" they moaned.

Wyatt wiggled his toes in the sunlight, trying to warm them.

Meg shivered. "Now that I've had my own personal shower, I'm cold. We'll head back to the car and warm it up. Hurry up, will you, Wyatt?" she said, nodding toward his socks and shoes still lying on the ground.

"All right, all right," Wyatt said. Benny grabbed Wyatt's sock and took off again, splashing his way across the creek to an island. "Benny!" Wyatt called, but the dog ignored him. He followed Benny across the shallow creek. Above him, a red-tailed hawk soared noiselessly. A belted kingfisher swooped by, his white necklace bright and his rattled call harsh. Dangling over the creek, an abandoned paper wasp nest hung, silent and still in the early winter afternoon.

Check It Out!

At the edge of the island, Benny sniffed at something. Wyatt climbed over a mound of dead plants, sticks and boards, debris left from former high waters, to get to the spot that Benny found so intriguing. A large pile of scat lay on the ground, folded neatly like a pretzel and crammed with tiny seeds.

> *Walking through the woods, your nose goes "ooh"*
> *You know some critter's scat's near you*
> *It may seem gross, well that's OK*
> *They don't have toilets to flush it away.*
> *Now don't go scream and lose your lunch*
> *If you look at it closely, you'll learn a bunch.*

The words to the art teacher's rap song seeped through Wyatt's mind as he poked at the scat with a stick.

> *It starts with an "S"*
> *Ends with a "T"*
> *Comes outta you*
> *And it comes outta me.*
> *I know what you're thinking.*
> *Don't call it that.*
> *Be scientific and call it scat.*

Rap It!

Clearly, this large animal had eaten birdseed, probably stealing from bird feeders out of either desperation or laziness. "C'mon, Benny," Wyatt called. "Let's get back to the car. I'm cold." He turned to climb back over the debris, glancing at Benny who rolled blissfully in the scat. "Benny," Wyatt scolded. *Why do dogs love scat so much?*

In the distance, Wyatt glimpsed a young woman picking late-season berries. Her long, dark braid and tan clothing explained why Wyatt hadn't noticed her earlier. *Camouflage*, he thought. She blended in well against the browns and grays of early winter. A sudden breeze made

Wyatt shiver. The woman reached into a small pouch, casting a powder onto the brambles. Three crows sat in a tree overhead, silently.

"C'mon, boy," Wyatt whispered to Benny, not wanting the woman to hear him. He carefully picked his way back across the creek. Benny heeled, uncharacteristically calm.

~ ~ ~ ~ ~

"There you are," Meg called. She stood beside the idling minivan, talking to someone Wyatt didn't recognize.

"We were getting ready to send out a search party," said Wyatt's father.

Wyatt grinned sheepishly and loaded Benny into the warm car. "Benny rolled in poop. Sorry."

"Did you wash him off before you put him in the car?" Meg asked, already sure of the answer. The man she chatted with chuckled.

When Wyatt shook his head, the man interjected. "I haven't shut off my outdoor hose yet, and I'm right around the corner. Why don't you come wash him off at my place." He directed Meg to his home adjacent to the Stroud Preserve. "I'll meet you there in five minutes," he added, hiking vigorously toward the bridge. "Then your car will just smell like wet dog instead of . . ."

"Scat," Wyatt said.

~ ~ ~ ~ ~

"Who was that guy?" Wyatt asked. They drove a short distance up Creek Road, crossing over a small tributary to the East Branch of the Brandywine.

"Hold on, Wyatt," Meg said. "Go left here," she directed her husband when they reached the intersection. "Over the bridge and take an immediate left. There it is." A sign at the entrance to the long drive read DEBORAH'S ROCK FARM and was immediately followed by another sign, PRIVATE; NO TRESPASSING.

"He was a guest speaker for one of my classes with Dr. Flo," Meg answered. "Mr. George works for Natural Lands Trust, an organization that works to save open space. Stroud Preserve is one of the properties they own."

The minivan climbed the hill to a peak overlooking the rolling hills of the Stroud Preserve. From the side of his house, Mr. George waved. He was holding a hose.

~ ~ ~ ~ ~

Once Benny was washed and back in the car, George's wife took Wyatt's parents on a tour of their home. George looked at Wyatt. "Your mom told me about your research. Good luck with that. I don't know if this will help with your mystery, but I'd like to show you something." He led Wyatt from the driveway to an outcrop of rock that jutted out and loomed a good twenty feet above the Brandywine.

"Wow," said Wyatt. He looked straight down to the water below. With the leaves off the trees, he could see Copes Bridge to the north and acres of fields and farmland to the east and south. Only a stone house and a post-and-rail fence interrupted the wheat-colored fields. A squiggly small creek entered the East Branch of the Brandywine from across the field. Wyatt realized that this was the small tributary they had driven over coming up Creek Road.

"What's that little stream?" he asked.

"That's Taylor Run," said George. "It starts out north of West Chester borough and runs into the East Branch of the Brandywine right over there." He pointed to the confluence just a few hundred yards downstream from where they stood. "The wastewater treatment plant is just up the run a bit."

Map It!

"Doesn't that pollute the Brandywine?" Wyatt asked, confused by the good Biotic Index he had gotten earlier.

"Not usually. They clean up that wastewater pretty well before they put it back into Taylor Run. But there are times when the wastewater isn't treated completely and that can be a problem."

"Like when it rains a lot. I know. My dad has to deal with that sometimes."

George hesitated. "He manages the Wilmington wastewater treatment plant, doesn't he? Well, then, you may already know about leaky pipes. So stop me if I'm boring you. See those houses up on the hill?" He pointed north. "Imagine all the pipes underground that lead from those houses to the wastewater treatment plant. Then imagine neighborhoods all over the area with underground pipes going to the plant. In some of the older neighborhoods, the pipes are pretty old and cracked, so groundwater can leak in and overload the sewer system."

Wyatt nodded, though he wasn't sure how this related.

"When it rains, the ground gets saturated," George continued. "Spaces in the ground fill with water. Water in the saturated ground needs to find places to go. If it finds holes in leaky pipes, it seeps in, sending lots more water than usual to the treatment plant. Since the plant is only able to clean up so much water at a time, it gets overloaded, and can send some of the wastewater through its system too fast and only partially treated."

"Like chocolate milk," Wyatt commented.

"That's putting it nicely," said George. They sat quietly for a few moments. The East Branch gurgled below while chickadees and nuthatches flitted among the tree branches.

"This is a really cool place," Wyatt said. "If I lived here, I'd come here all the time."

"Wyatt's Rock. It's got a nice ring to it. Unfortunately for you, it's already got a name: Deborah's Rock. This farm is named after it."

"Who's Deborah?"

"There seems to be some conflicting stories about that," George said. "She may have been a member of the family who originally owned the farm. Or she may have been a Lenape woman. We'll never know for sure."

As the pair retraced their steps to the driveway, Wyatt spotted a seedy mound of scat similar to the one Benny found on the island.

"Coyote," said George.

Check It Out!

~ ~ ~ ~ ~

The sun had set and it was cold by the time the Nystroms drove into West Chester to meet Danni's family. The streets were magical. Draped with pine greens and red bows, the streetlights sparkled. Carolers sang from street corners, bells rang in front of festive storefronts and people crowded the street. The odor of popcorn and hot chocolate drifted through the air. On the sidewalks, people greeted each other. Everyone seemed to be in a good mood.

The Nystrom clan entered the Chester County Historical Society for their favorite part of the Old Fashioned Christmas celebration, the storyteller. Wyatt and Danni laid their jackets on the floor to reserve spots close to the stage. Then the family wandered out to the museum to kill time before the performance. The women headed to the quilt display that Wyatt's mother and aunt wanted to see. The men went to an adjoining room that featured relics of famous visitors to Chester County like the Marquis de Lafayette, who fought with George Washington during the Revolutionary War. Abraham Lincoln, whose biography was published in West Chester, had passed through the county by train several times during the Civil War. "Considered by some to be a time of madness," Wyatt read. *More madness*, he thought.

Check It Out!

Wyatt followed Danni into the quilt room, where she stopped in front of a pretty, yet unremarkable, quilt. FLYING GEESE read the sign beside the quilt, sewn with blue and yellow fabric blocks.

"THOUGH HIGHLY DISPUTED BY HISTORIANS, FOLKLORE EXISTS CLAIMING THAT QUILT PATTERNS WERE USED AS PART OF THE UNDERGROUND RAILROAD PRIOR TO THE CIVIL WAR. THE 'FLYING GEESE' PATTERN ON THIS QUILT WOULD HAVE PROVIDED INSTRUCTION FOR RUNAWAY SLAVES TO HEAD NORTH IN THE SPRING, IN THE DIRECTION OF MIGRATING GEESE, TO FIND FOOD, WATER, AND FREEDOM. THOSE THAT DISPUTE THE QUILT THEORY MAINTAIN THAT ESCAPING SLAVES COULD HEAD NORTH SIMPLY BY FOLLOWING THE NORTH STAR IN THE NIGHT SKY."

Check It Out!

"Mom, look," Danni said to her mother, who was reading the description. "Why would they have these quilts in Chester County? Wasn't Pennsylvania a free state?"

"Yes, honey, prior to the Civil War, Pennsylvania was a free state, so many runaway slaves left places like Maryland and Delaware and headed for Philadelphia. But then Congress passed the Fugitive Slave Act that allowed bounty hunters to capture escaped slaves in free states and return them to their slave owners. So the Underground Railroad continued through the free states to Canada, where slavery was completely outlawed."

"It wasn't a real railroad, was it?" Wyatt said. "They just used the idea of a railroad, right?"

"Yes," June answered. "The *railroad* was just the term they used for the way the slaves traveled through forests and along waterways, or hidden in wagons. *Stations* were homes used to hide and shelter the escaping slaves and *conductors* guided them. Since there were so many free blacks and sympathetic Quakers here, Chester County was an important part of the Underground Railroad."

Check It Out!

An announcement by a museum guide interrupted their conversation. "The storytelling program is about to begin. Please proceed to the auditorium to take your seats before the lights are turned out." Luke and John Nystrom emerged from the back room to accompany their wives and children back to the auditorium.

Danni and Wyatt sat side-by-side on their jackets in the dark room, now filled with seated kids. Most of the parents stood along the surrounding walls. The darkness had a quieting effect on the crowd, aside from some whispering and an occasional cough. A single candle rested on a stool on the stage, and a man dressed completely in black emerged from behind a curtain. He sat behind the candle that illuminated his pale face and blue eyes.

"Good evening," he began in a serious and slowly-paced tone. "And welcome to an Old-Fashioned Christmas story hour. I'm Bill, and I'll be guiding you through a few stories based on legends. As the years go by, these legends have taken on new dimensions, so no one really knows what is true and what is made up. It is for the listener to decide. Before I get started, let me tell you that if you should fear scary dreams tonight after hearing any of these stories, cross your shoes under the place you lay; this will keep bad dreams at bay." A few cries from the audience erupted but were quickly quelled with bribes of hot chocolate and candy canes. Several parents seized their children by the armpits to escort them to some jollier activity. Wyatt and Danni squirmed in their seats as the scuffling died down. They both loved creepy stories.

"A long, long, long time ago," Bill began, and he told the Lenape story of how crow, once the colors of the rainbow, came to be black with a hoarse, ugly song and how his black color and unpleasant voice has kept him the most free of all animals.

Check It Out!

The next story was adapted from a Navajo legend about how the stars got placed in the sky. "The stars were supposed to be placed thoughtfully, to represent the laws of the people. But coyote was impatient. He tossed them into the sky without care, just to get the job finished," said Bill. "And that left the people forever in confusion."

Check It Out!

It seemed like a weird coincidence that Bill told a story about a coyote when just a few hours earlier, Wyatt had learned that coyote lived in the area. And that the other story was about crows; he had been noticing crows a lot lately.

The candle in front of Bill was decorated with puddles of melted wax. "For my last story this evening," Bill continued, "I will tell my story of Deborah's Rock which, I must confide in you, is not accepted by all. Some say Deborah was simply a farm girl who enjoyed the view from her special rock overhanging the Brandywine. Others believe Deborah was a victim of circumstance. Listen to my tale and you be the judge."

"I was just there," Wyatt whispered to Danni. "At Deborah's Rock. I swear!"

Danni looked at Wyatt skeptically and turned her attention back to the storyteller.

"For many centuries, the Lenape Indians lived in this area. They migrated along the Brandywine Creek, hunting in the woodlands, fishing in the streams, and harvesting corn in the growing season. They lived with, and from, the land, moving as the seasons dictated, taking only what they needed and what the Earth provided. Then," Bill paused, shaking his head slowly. "In the sixteen hundreds, the Europeans arrived. At first, they came for beaver to ship back to Europe to make fine hats." Bill reached behind the stool and placed a fur hat on his head. The candle flickered in the slight breeze he created. "They settled in places along the Brandywine, where the water was sweet and powerful, the forests were crowded with trees, and the soil was fertile. Immediately, they began clearing those forests for farmland and building gristmills for grinding their wheat into flour. Over the course of a few decades, the Europeans changed the landscape, forever altering the Lenape's way of life. Dams slowed the creeks, blocking shad from migrating upstream to breed, and changing roaring waters to tame currents. Woodlands turned to farmland, and hunting waned. Lenape encampments vacated during the winter season were claimed and settled by the Europeans, leaving the Lenape homeless upon their return in the spring. Many Lenape left the area to seek new lands. Some stayed behind, including Deborah."

Check It Out!

Bill paused. He took a drink of water from a glass on the floor and then continued his story. "Deborah remained behind with others in her tribe, trying to live off the land as they had always done. But now, instead of other Indians living in camps along the trails, white settlers had claimed properties. Not understanding or valuing the concept of land ownership, the Lenape often wandered through these properties.

"One late afternoon, Deborah was headed to the creek from the woodland where she and others of her tribe had spent the day tanning hides and grinding corn. Dressed in deer hides, her long black braid swinging across her back, she carried her gourds toward the creek to gather water for the evening meal. By the white man's house, she saw a young man halt his wood chopping. She looked to see him standing still, watching her. Unlike the exciting sensation she felt when observed by young men of her tribe, she felt an unpleasant, eerie sensation in her belly from this white man leering.

"Day after day, the white man followed Deborah with his eyes. She changed the path she took to the creek, but somehow, the man was always nearby, staring. Deborah asked a friend one day to accompany her, but at the last minute, the friend's mother summoned her to another task and Deborah had to make the trip to the creek alone.

"Deborah took the original path to the water right by the white man's home. The young man was nowhere to be seen. Deborah felt better, but was still cautious. She continued down the trail, taking note of the sun sinking early as the fall settled in. She smiled at the flock of geese flying overhead, honking as they went on their migratory journey. Chipmunks scampered by, nuts in their cheeks. Crows cawed as she passed underneath their perches. The leaves rustled gently in the breeze."

Wyatt sat still. Danni leaned against him, her hand clutching his left arm. Bill continued. "A sudden movement behind a poplar tree caught Deborah's eye. Knowing full well that coyotes and bears and turkeys don't bother to hide behind trees, she felt a tightening in the pit of her stomach. She closed her eyes and whispered a quick prayer to Ketanetuwit, the Great Spirit. Then she reached into her pouch and blew a powder made of her sacred herbs in the direction of the tree. Suddenly, the white man stepped out from behind the tree with a sinister smile across his face. His front teeth were stained and twisted and his dark hair stuck out in all directions. Sweat trickled down his face as he lunged at Deborah." Bill stopped. He drank slowly from his glass. The auditorium was silent. The candle flickered.

"It is said that a young, eligible woman must be untouched by a man until her wedding night. Deborah ran. She fled down a path as the man gave chase, gaining on her. She stopped. The path ended abruptly at a cliff above the shallow Brandywine. She turned, and the man charged. Deborah chose her destiny."

Bill stopped speaking, lowered his head and sat very still. Moments stretched to a full minute. The audience became fidgety. Whispers rose. "What happened to her?" "What did she do?" "What does *destiny* mean?"

Slowly, Bill raised his head, and stared directly into the candle. "Deborah hurled herself from the cliff," he said, "plunging to the deadly stream below." He blew out the candle and the room went completely dark. Girls screamed and parents hushed them as Bill continued to speak in a low but firm voice. "It is said that if you walk the grounds near Deborah's Rock, you may see the ghost of Deborah to this day, trying to live the Lenape way."

~ ~ ~ ~ ~

Wyatt turned out his bedside lamp and closed his eyes. He reviewed the evening's festivities in his mind: the sprinkling of snowfall during the Christmas parade; the sparkling lights of the town-center tree; the chocolate peppermints tossed by Santa as he passed through the streets of West Chester in his horse-drawn sleigh; the storyteller's tale of Deborah. He sat up and snapped on the lamp. Pushing his comforter aside, he rose from the bed. Thoughtlessly tossed into the back of his closet were his sneakers. He retrieved the shoes and placed them under his bed, crossing them.

Field Trip!

Chapter 16

ROYAL FLUSH

A large manila envelope lay on the kitchen table. Scrawled in large black letters across the front was Wyatt's name. He pulled out the papers.

"We hope you can use the data we collected. Please let us know what you find out. From Mr. Maxwell, 5th Grade Teacher, Hillsdale Elementary School."

"I see you found the envelope from Dr. Flo," Meg said to her son as she entered the kitchen. "She gave it to me the other night after class. It looks like you got some more participants. Is there anything good?"

Wyatt fanned through the set of papers. "It looks like there's data from two schools in West Chester. It looks like they tested together. And there are a few others, too."

"Great," his mother said. She grabbed a tub of cream cheese from the refrigerator. "How about a bagel? Are you hungry?"

"Sure," Wyatt said, "if you're making it. Thanks." He sat at the table reading while his mother puttered around the kitchen, filling the coffeepot from the tap, pouring glasses of orange juice, toasting bagels, and setting the table around him.

Date: November 12
Time: 1pm
Location: East Branch of Brandywine above and below confluence with Taylor Run; at Stroud Preserve

Hey, that's where we were last week, he thought. He tried to remember what unofficial Biotic Index he had come up with before his feet had gotten too cold. *I hope these kids wore boots*, he thought, but then realized they had done their test almost a full month before he had. He read through the rest of the data.

Method of Sampling: general area search
Weather Conditions: sunny, about 55 degrees

Weather in Last 24 Hours: same
Biotic Index: Above confluence with Taylor Run =8
Biotic Index: Below confluence with Taylor Run =11
Points of Interest: Taylor Run (upstream is wastewater treatment plant—see attached); Georgia Farm with riparian buffer being grown
Submitted by: Hillsdale and East Bradford Elementary Schools

Wyatt hadn't heard the term riparian buffer before, and was impressed by these fifth-graders. "What's a riparian buffer?" Wyatt asked his dad, who entered the kitchen.

"I smell something good," Luke said, giving his wife a hug. Wyatt rolled his eyes. "A riparian buffer? Riparian means along a river. And a buffer is a form of protection. Riparian buffers are trees and shrubs planted along streams to protect them from runoff."

"Is that what those thingies that looked like birthday candles were at Stroud Preserve? The ones in the fields near the creek?"

"Indeed they are. Since, historically, people haven't understood the importance of trees and shrubs along creeks to protect water quality, they've mowed and plowed right to the banks of a creek. What that does is take away the river's defense from all the crud that can wash off the land into it. Now many conservation groups are planting riparian buffers along creeks. Unfortunately, deer really like to eat the seedlings they plant, so people protect the seedlings with plastic tubes until the baby trees are large enough to survive on their own."

"What kind of crud? Like pesticides and fertilizers?" Wyatt thought back to the powdered drink packets at Dr. Flo's house.

"Yes, and soil and manure and all sorts of things. A line of trees and shrubs helps to stop that stuff from going directly into the creek. The buffer slows down the runoff so it can infiltrate into the ground. Some plants actually absorb some of the pollutants; other pollutants get filtered out by rocks and dirt. Here, let me show you something." Luke beamed as he instructed his son in an area related to his work. He handed Wyatt a clear glass. "Fill this and put in some food coloring." Luke then pulled some leafy stalks of celery from the refrigerator and angled the bottoms with a knife. He placed them in the glass of blue-colored water with the leaves sticking out above the rim. "Give it a few hours and then check what happens. It'll show you what some plants do with pollution." He hesitated before continuing. "Actually, riparian buffers act similar to wastewater treatment plants. They both filter out wastes before re-supplying the stream. And why do you think they're both called 'plants'?" He winked.

Try It!

"Bagels are ready," Meg said. "So, Wyatt, what are you up to today?"

Wyatt thought for a minute. "I think I'll see if Danni is around and maybe we'll drop by Dr. Flo's to put this new data on the map." Wyatt scanned the note from Mr. Maxwell's class. "Hey, Dad, look at this." He showed the paper to his father, who looked it over.

"Not a bad description," he said. "Similar to the Wilmington plant; we just use different technology."

Wyatt took the paper back and read silently. Apparently this was a student's extra credit paper that the teacher thought worth forwarding to Wyatt.

> "Taylor Run Wastewater Treatment Plant: A Brief Description: Today our class went for a tour of the Taylor Run Wastewater Treatment Plant in West Chester. It stunk! But the field trip was good. The TRWWTP treats wastewater from homes and businesses in West Chester Borough and East Bradford Township, including our school. They treat about one and a half million gallons of wastewater each day and more on Super Bowl Sunday (the guy said there are lots of toilets being flushed that day)."

Wyatt recalled the last Super Bowl party at his home; the downstairs bathroom was constantly occupied, so a lot of guests ended up using his bathroom upstairs. And some of those people were fat. And greasy from eating chicken wings.

"The treated wastewater goes back into Taylor Run and goes to the East Branch of the Brandywine Creek to the Christina River to the Delaware River to the Delaware Bay and finally to the Atlantic Ocean. So they have to put the treated wastewater back pretty clean because other towns use the water downstream."

Yeah, like me, thought Wyatt, *and the rest of us down here in Wilmington.*

Map It!

"There are many laws and rules about how clean the water has to be before it can be put back, but they are discovering some problems with all the different things people flush down their toilets and put down their sinks. Lots of stuff from people's homes are not even tested for, so some things are getting into our water that we don't even know about (like medicines and chemicals). This can cause problems for fish and frogs and other wildlife, though it's not a problem for people . . . yet."

"Dad, is this true?" Wyatt asked. "The part about chemicals that we don't even test for?"

"Afraid so, son. There are so many products people use in their homes now, and many of them are full of chemicals. We're not sure what they all are, so we're certainly not sure how to manage them all." Luke hesitated. "You know, Wyatt, I'm really proud of this project you're doing. Water quality is really important. It's what everyone needs to live, but also what everyone takes for granted."

Wyatt grinned sheepishly. His mom smiled as she poured the coffee. "My two boys talking shop. It warms my heart, even if it is talk of . . ."

"Scat!" blurted Wyatt. Meg and Luke laughed.
Wyatt continued to read to himself.

"The first thing we saw was a big tank
with a screen. This is where the wastewater
comes into the plant in pipes from houses,
businesses and schools. In the tank, big stuff
gets screened out from the wastewater,
including lots of used toilet paper. The guy
told us that once they found a hardhat in the
tank and hoped the person wearing it didn't
show up next (he didn't). The water here was
brown and bubbly and smelly! Besides the
big stuff that gets screened out here, heavy
stuff that sinks gets taken out.

"Next, the wastewater goes through these
really big white rotating tanks. The insides
of these things are totally gross and slimy
because they are covered with bacteria that
use oxygen to 'eat' the waste. The bacteria
get heavy and sink down and then get taken
out (this is called sludge). The sludge gets
dried out and sent to the landfill, where all
of our trash is taken.

"The wastewater goes through a bunch of
these tanks and gets cleaner after each one
until it is finally clean enough to put back
into the creek. They add chlorine and then
send the treated wastewater back into Taylor
Run. For all of this to happen, it takes about
a day or two. It goes into the East Branch
of the Brandywine just downstream from
Ingram's Mill. The guy said it's very important
to send your treated wastewater back below
the place you get your drinking water."

But it's still upstream from my drinking water, thought Wyatt.

Check It Out!

"Dad, is the Wilmington Wastewater Treatment Plant downstream from where we get our water?"

"Actually, Wyatt, Wilmington gets its drinking water from the Brandywine, but we put our effluent into Shellpot Creek and that goes to the Delaware River. Wilmington crosses watersheds to dispose of its treated wastewater."

"That's good," Wyatt said. "At least for us."

"We're still downstream from other communities' wastewater treatment plants, like Taylor Run. And two in Coatesville. And one in Downingtown. We all live downstream," said Luke.

Wyatt watched his dad stir cream into his coffee. He shuddered at the sight of the swirling light-brown liquid. *Now that looks like wastewater,* he thought.

~ ~ ~ ~ ~

At Dr. Flo's, Wyatt pressed a red sticker on the wall map on the East Branch of the Brandywine just above the confluence with Taylor Run. He pressed a blue sticker a little way below the confluence. He traced Taylor Run upstream to the wastewater treatment plant in West Chester.

Red Stickers	Blue Stickers
East Branch Brandywine above Taylor Run; BI=8	East Branch Brandywine below Taylor Run; BI=11

Sticker It Red or Blue!

Dr. Flo entered the room. "So, what do you see so far?" she asked.

"The water quality is good in some places, but not so great in others," said Wyatt, looking at all of the sites tested.

"Is there a pattern?"

"It's hard to tell," said Danni. "Some of the tributaries are worse than the Brandywine."

"Interesting, Danni. Why do you think?"

Danni and Wyatt looked at each other, shrugging. Dr. Flo motioned for them to follow her into the kitchen. She retrieved a small shot glass, a clear pitcher filled with water and a bottle of red food coloring. "Put five drops of the food coloring in the shot glass. Then fill it with water," she said.

Danni followed the instructions. The water in the shot glass turned scarlet.

"Now pour that into the pitcher of water."

A plume of scarlet hit the water and migrated throughout the pitcher. The water turned the color of watermelon.

"It got lighter," Danni said.

Try It!

Dr. Flo led the children back to the map. "Look at the Biotic Indexes for Radley Run and Plum Run. Then look at the Brandywine."

Like a fluorescent bulb warming to its full brightness, Wyatt took a few seconds and then completely understood. "The tributaries aren't that clean, but when they go into the Brandywine, there's a lot more water, so the effect isn't as big as it is in the smaller tributaries."

"Exactly," said Dr. Flo. "Back in the olden days, they often said, 'The solution to pollution is dilution.'"

"Yeah, but the Brandywine near these tributaries has a lower Biotic Index than farther upstream, so the Radley and Plum Runs must have some effect on the Brandywine."

"I never said we were right in the olden days." Dr. Flo laughed. "The added water dilutes the effects, it doesn't remedy them."

Danni looked at the stickers in Delaware. "Then how come the water quality isn't good in Wilmington if that's where the most water is? Wouldn't it be really diluted there, since there are so many little streams adding water to the Brandywine?"

"Ah, Danni, you are also correct. The amount of water flowing through Wilmington is far greater than anywhere else on the Brandywine. The quality of the Brandywine is an interplay between the quantity of water and the quality of its tributaries."

"So, for the Brandywine to be clean, the tributaries have to be clean, right?"

"Right," Dr. Flo said. "If there are problems upstream, they will accumulate downstream, in spite of how much water there is."

"Like the spit!" Wyatt remembered the example from his first real lesson about water quality with Dr. Flo at Alapocas Woods.

Dr. Flo smiled. "Like the spit. It may dissolve into the water, but it accumulates downstream with all the other spit."

"I said I was sorry about that," Wyatt said, pretending to be offended.

Dr. Flo winked. She examined the stickers around Taylor Run. "What did the data sheet say about the points of interest here?"

Wyatt summarized the student's description of the wastewater treatment plant. He purposely included the slimy bacteria just to get a rise out of Danni.

"They must be doing a pretty good job at the plant, since there is good water quality below the confluence with Taylor Run," remarked Dr. Flo.

"My dad says that the birthday candles help."

Dr. Flo raised a questioning eyebrow. Wyatt explained the concept of the riparian buffer and the way the seedlings looked like birthday candles with their little limbs sticking out the top of the plastic tubing like wicks.

"I don't get it," Danni remarked. "How do the trees stop pollution?"

"Demonstration time," Dr. Flo sang out. She led them outside. "I think we can still get away with this, since the ground isn't frozen yet." She handed Danni a spade, directing her to dig some dirt and mix it into a nearby bucket of water. She then led the children over to a large plastic barrel with a clear viewing panel running from the top of the barrel to the bottom. The barrel sat atop two concrete blocks with space between them. From the top of the barrel, Danni could see a deposit of grass, weeds, and dead leaves above a layer of soil. Below the soil was a layer of sand followed by small rocks evolving to larger rocks. "This soil filter is like the ground, with all the layers underneath us that we can't see. Now, take the dirty water and pour it into the barrel and watch what happens." Danni followed the instructions.

Try It!

Water began to drip slowly from the bottom of the barrel where there was a small hole. Dr. Flo placed a clear plastic cup under the filter to catch the dripping water. "Remember the water going in?" Wyatt and Danni nodded. She pulled the cup out. "Ta da!" she said, producing a cup filled with crystal clear water.

"That's like magic," said Danni. "How did you do that? I thought the water would just get dirtier."

"That's the beauty of Mother Nature," said Dr. Flo. "The ground acts like a tremendous wastewater treatment facility, filtering out pollutants and dirt, and *cleaning* the water. But it can only do its job if there are trees and other plants to slow the water down so it can get absorbed into the ground. The more trees and shrubs we have on the ground, and the less pavement, the more water absorbs into the ground instead of running off the surface. And the more absorption into the ground, the cleaner the water becomes before it is recharged back into the stream."

Try It!

"Isn't that what the wetlands do?" said Danni.

"Yes, woodlands and wetlands are both great for cleaning water."

"So that would explain why the there are higher Biotic Indexes around places like the Stroud Preserve and Brandywine Creek State Park, right?" said Wyatt. "Because the creek is surrounded by woods."

"Indeed," said Dr. Flo.

"Well, I guess this is all starting to make sense," said Wyatt.

~ ~ ~ ~ ~

After a cup of hot chocolate, Danni and Wyatt said goodbye to Dr. Flo. "No ride today?" she asked.

"No, we'll just walk. It's not that far."

On his way to the front door, Wyatt looked at the portrait of Uncle Clayt. The dignified man looked less formidable with his slight wink and shadow of a smile. *Was he nodding before?* Wyatt thought. *Must have been.*

Danni skipped along the sidewalk, trying to warm up. The air had turned colder and the sky had clouded over. The mile-long walk to Wyatt's home was chilly, but gave him a chance to observe just how much of Wilmington was built up with homes and buildings and paved over with driveways and streets.

A sudden downpour came from nowhere. Wyatt and Danni ran to take cover under an evergreen tree. A few minutes later, the rain stopped as suddenly as it had started. Soaked, Wyatt and Danni continued to walk home. They crossed Riddle Avenue where a brown streamlet, bounded by the curb, gushed down the storm drain with an echo. *That dirty water came from all of these properties,* thought Wyatt. *And it all goes straight to the creek without being filtered through the ground first.*

Check It Out!

~ ~ ~ ~ ~

"You're soaking!" Meg Nystrom said when the two children walked through the front door. "Don't move." She disappeared up the stairs and returned with two towels. "Go change into something dry," she ordered, tossing the towels at them.

They peeled off their wet socks and shoes and trudged across the kitchen. Wyatt looked up to the counter where the celery stalk stood in the glass of blue water. "Look," he said. Danni sidled up next to him and peered at the celery stalk. Its leaves were blue.

Field Trip!

Chapter 17

LONGEST NIGHT OF THE YEAR

The weeks before Christmas were packed with activities for the Nystroms: winter concerts, holiday fairs, office parties and shopping, shopping, and more shopping. Although it was a busy time, the two cousins and their families always celebrated their favorite event of the year, the winter solstice.

Wyatt and Danni strung popcorn and pieces of fruit for the celebration. When they were younger, they believed the folklore that animals of the world could talk on the longest night of the year. To continue the tradition, they suspended common sense and made edible treats to drape on the evergreen trees for the animals' all-night party.

Try It!

"Are you finished?" Meg asked, entering the room with Danni's mom and dad. "We've got to get going."

The kids loaded the garlands in a plastic bag and gathered the blankets, extra clothing, and flashlights. Even though the celebration at the ChesLen property didn't begin until sundown, Wyatt knew there were errands to run beforehand, up in Pennsylvania territory.

Packed tightly into the car, the two Nystrom families, along with Benny, rode up Route 52. They passed the entrance to Winterthur, cruised slowly through the quaint village of Centerville, and crossed into the state of Pennsylvania, just minutes from their home in Delaware.

Map It!

They drove through Unionville until they spotted signs for Stargazer's Winery and Vineyard, then crossed a bridge to a winding, narrow road that paralleled the creek and a set of railroad tracks. A train approached, carrying several huge tanks, and blew its whistle. LIQUID PETROLEUM GAS was printed in bold letters across the tanks.

"Are those the same train tracks that we saw near Pocopson School?" Wyatt asked.

"Here, check it out." Luke handed his son a road map of Chester County, Pennsylvania. Wyatt traced the tracks running along the Brandywine that connected Wilmington to Coatesville.

Map It!

"In case you were wondering, yes it is," Wyatt informed everyone.

The car continued partway up a hill to a driveway at the end of a lane. As the family clamored out of the car, a pair of small dogs ran toward them, yapping and jumping. Benny ran directly to the dogs that set about shamelessly sniffing each other's privates. The humans turned away to take in the view of the landscape, rather than the rear ends of the dogs. The winery, with its black solar panels absorbing the weak afternoon sun, faced a landscape of forested hills. Even in winter, the setting was striking.

"Hello, Nystroms," a man greeted. "Glad you made it."

Luke and John shook hands with their old high school friend, then introduced their wives and children. "You kids can play with the dogs if you want, while I set your folks up with our solstice special."

"Why is this place called Stargazer's Winery?" Wyatt said. "Is it because of the view?"

"Well, it could be due to that. It is a heck of a view. But more likely, it's because we're just a hop, skip, and jump away from Stargazer's Stone."

"What's that?"

"It's the observation site used by the astronomers Mason and Dixon, to determine the boundary between Pennsylvania and Maryland."

"Why didn't they just use a map?" said Danni.

"Maps weren't very accurate back in the seventeen hundreds," the man explained. "People were arguing over property boundaries, so the English commissioned Mason and Dixon to do a survey that everyone could accept. Have you ever heard of the Mason-Dixon line?"

"Yeah, it has something to do with slavery," said Wyatt.

"Exactly. The Mason-Dixon line determined the boundary between Pennsylvania, a northern state, and Maryland, a southern state. So it was pretty important to know exactly where that boundary was, because it separated the free states from the slave states."

"Yeah, until the Fugitive Slave Act," added Wyatt.

"You must be a history buff," said the man. "Pretty impressive. Now, I've got some impressing to do myself, with my wine. Nice chatting with you."

Check It Out!

~ ~ ~ ~ ~

The drive to Highland Orchard should have been a short one, but was interrupted when Wyatt saw the road sign marked STARGAZERS. At his request, his dad turned onto it and drove north slowly.

"That must be it," Wyatt called, pointing to a rock structure on the sloping yard of an old house. "That must be where the rock is that they used to figure out the boundary between the north states and the south states before the Civil War."

Map It!

"Mason and Dixon?" Meg asked. "How do you know that?"

"That guy from the winery was telling me and Danni about it," he answered. "Mason and Dixon used astronomy to do the measurements to figure out exactly where the boundary between Maryland and Pennsylvania is."

"You're a regular encyclopedia, Wyatt," commented Uncle John. He turned to Wyatt's father. "Do kids even know what an encyclopedia is?"

They continued to drive past the entrance to the Embreeville Center, turning left onto Marshalton-Thorndale Road, and began an uphill climb. The car strained as it headed up. And up. And up. When they reached the top of the hill, fields and sky opened up. They pulled into the orchard's parking lot, where the marketplace advertised homemade pies.

~ ~ ~ ~ ~

"Oh, I get it," Wyatt said. Danni swung on a tire swing as Wyatt reached over the fence to pet the coarse, white fur of one of the orchard's resident goats. Benny whined from inside the car, clearly wanting to sniff the new species. The adult Nystroms had gone into the marketplace, leaving Danni and Wyatt to keep occupied with the various kid-friendly attractions. They had been coming to this market for as long as Wyatt could remember, but it never occurred to him to question the origin of the orchard's name. "Look at the name on the sign, Danni."

"So what?" Danni said. "It's the same name it is every year, Wyatt. 'Highland Orchard.'"

"Did you notice the hill we came up to get here?"

"Yeah. So?"

"This orchard is in the *high land*. Get it? *Highland Orchard*," he explained.

"Oh, yeah, I get it," Danni said.

Wyatt ran to the car to examine the Chester County map, looking for the spot where they now were. *This must be the place*, he thought, as he located a grid of small lanes with names like Macintosh and Winesap. Recalling his early lesson with Dr. Flo about high points dividing watersheds, he began tracing several nearby blue lines to their major creeks. "Look, Danni," he pointed out. "Everything on this side of the road goes to the East Branch of the Brandywine."

Danni traced the blue lines on the west side of Marshalton-Thorndale Road. "And these little streams all go to . . ." Her voiced trailed off as she looked for the name of the large tributary where the streams flowed. "Broad Run which goes to the West Branch of the Brandywine Creek. And look," she said, tracking the big blue line further. "Here's where the West Branch meets the East Branch of the Brandywine."

Map It!

Wyatt checked and, sure enough, the East and West Branches of the Brandywine merged together not far below Broad Run's merge with the West Branch. "I wonder if this orchard sprays their apples with pesticides," Wyatt said, now knowing how likely it would be for those substances to end up in the water supply downstream.

"Or uses fertilizers," Danni added. "Why don't we ask this school to do some tests?" She pointed out West Bradford Elementary School on the map, located on Broad Run.

"That's a great idea," Wyatt said. "And we can get this school to test on the East Branch." He pointed to an elementary school near the East Branch called Bradford Heights.

~ ~ ~ ~ ~

The smell of freshly baked apple crisp wafted through the car. Wyatt's stomach grumbled, but he knew not to ask for a piece of pie before the solstice celebration potluck. Trying not to think about food, he concentrated on the sites they passed. In the village of Marshalton, a sign with four Labrador retrievers hung outside the Four Dogs Tavern. "Look, Benny," Wyatt said. "Relatives." At the historic inn next door, he imagined stagecoaches lined up and the inn full of fancy people while, down the road, simply-dressed Quakers met at the Bradford Meetinghouse. He peered at the electric candles in the blacksmith's shop and pictured the likely businesses of the 1700s: the cigar maker, the shoemaker, the tinsmith. He smiled at the colorful WELCOME and SEASONS GREETINGS flags draped from the restored village homes. On the far side of town was MARTIN'S TAVERN, EST. 1764. As Luke slowed the car at the curve, Wyatt was able to read a sign describing this site serving Washington's men during the Revolutionary War and possibly even Mason and Dixon.

Map It!

They drove downhill to Embreeville, passing soccer fields on the right just before the entrance to EVENT PARKING. Luke followed the instructions of a man directing traffic and parked up the drive beside a NO TRESPASSING sign.

Wyatt and Danni got out of the car. "If I had to take one more whiff of that pie without being able to eat it, I would have died," Wyatt whispered. Benny dashed out and ran over to a huge boulder, set back from the road and oddly out of place. As their parents unloaded the gear, Wyatt and Danni ran toward the dog, hoping to leash him before he realized he was free on this vast property.

"Yuck," Danni cried, as Benny lifted his leg.

Wyatt rounded the white quartz boulder to leash Benny and stopped. There was a bronze plaque mounted on the boulder.

HERE RESTS INDIAN HANNAH,
THE LAST OF THE LENNI-LENAPE INDIANS IN CHESTER COUNTY
WHO DIED IN 1802.

Map It!

"Danni, look," he said, grabbing Benny's collar and securing the leash on him.

"Indian Hannah? The one with the marker near Longwood Gardens? Why would she be buried here? It's not even a cemetery."

Wyatt shrugged his shoulders. "I have no idea," he said.

The family joined a line of people carrying blankets and baskets of food and walking carefully along the side of the road. Cars cruised by, slowed by the bright-orange cones in the middle of the road and the men in reflective vests directing traffic.

The Nystroms crossed the road and walked over the old stone bridge above the West Branch of the Brandywine. At the sight of the open fields and wetlands of the ChesLen property, Benny strained at his leash. "I wish I could let you run around, Benny, but there are too many people," said Wyatt.

Map It!

A trail of solar lights marked a route straight ahead. The Nystrom clan followed the crowd across a set of train tracks. In the distance, a train whistled. Meg's cell phone rang.

"Hello?" she answered. "Hi, Flo. Yes, we're almost there. We've just crossed the train tracks . . . OK . . . OK. What is it? OK, we'll see you in a few minutes."

Meg turned to her family. "That was Dr. Flo. She's waiting for us at the campfire site. She says they won't be starting the celebration for a little, so we should stop at Potter's Field."

"What's that?" said her husband.

"She wouldn't say. But she did insist that we check it out. She said that it'll add some *spirit* to the evening."

Ahead of them, a group of people stood clustered around a wooden sign posted beside a fenced-in yard. Words were burned into the wood: POTTER'S FIELD EST. 1800; KNOWN BUT TO GOD, PROTECTED BY US. Someone opened the unlocked gate of the fence and a few people entered the yard.

"C'mon," Wyatt urged the adults as he and Danni followed the others toward the far corner of the yard. Spaced at regular intervals, small rectangles of polished granite were half buried in the ground. Each one was blank except for a number etched into the center. Plastic daisies decorated marker "1." Meg, Luke, June and John approached.

"Wow," said Meg. "Anonymous graves." The adults stood silently, reverently.

Danni and Wyatt exchanged glances. "You mean this is a graveyard?" said Danni. "With dead bodies?"

"Whoooooooo," Wyatt moaned. "Daaaaaannnnnnniiiiii," he teased.

"Cut it out." Danni slapped Wyatt lightly.

"Apparently, these were residents of the Chester County Poorhouse who had no family to bury them when they died. So they were buried here," said June.

"And only given numbers," Meg said. "No names. You don't even know whose grave is whose."

Danni thought for a moment. "And Indian Hannah was buried near where we parked," she said. "There sure are a lot of dead people around here."

Oh great, thought Wyatt. *Sometimes, I see dead people.*

~ ~ ~ ~ ~

The family settled onto a thick blanket as the darkness of the shortest day of the year enveloped them. Bundled in layers of sweatshirts, jackets, mittens, and hats, they sat still for the moment of silence opening the evening ceremony. Dr. Flo stroked the back of Benny's neck to keep him quiet.

A woman appeared beside an unlit pyre and bowed toward the audience. She was dressed in red regalia and her long, dark hair was decorated with feathers and beads. She bent down with a lighter. From the center of the pyre, crackles sounded as the fire caught hold of the dry wood. Flames rose, gradually at first, then danced their way toward the top. Sparks shot into the air, burning out quickly like little meteors.

"*Lomewa, luwe na okwes xu laxakwihele xkwithakamika,*" the woman began in a low monotone. "Long ago it was said that a fox will be loosened on the earth." She told her tale in the strange foreign tongue, translating to English after each sentence. "Also it was said four crows will come. The first crow flew the way of harmony with Creator. The second crow tried to clean the world, but he became sick and he died. The third crow saw his dead brother and he hid. The fourth crow flew the way of harmony again with Creator. Caretakers they will live together on the earth." She lowered her head and quietly walked away.

"Thank you, Lee." A man's voice boomed as he stepped in front of the now blazing bonfire. Wyatt recognized Mr. George from the Stroud Preserve, the man who took him out to Deborah's Rock. "And good evening, everyone. Welcome, once again, to our annual Winter Solstice Celebration. As many of you already know, I'm George, from Natural Lands Trust, and you are at the ChesLen Preserve, the largest private nature preserve in southeastern Pennsylvania." He spent a few minutes explaining the local history, including the area where the Nystroms had parked the car. Presently owned by the state of Pennsylvania and housing PennDot and a school, the property had been the site of the Chester County Poorhouse in the 1800s and *a Lunatic Asylum.*

"Which we now more sensitively call a Mental Health Facility," he added. "You may have noticed the Potter's Field on your way here. It is the burial site of hundreds of anonymous residents of the Poorhouse. The most famous resident, of course, was Indian Hannah, who spent her final years at the Poorhouse. She was honored with a monument and a special burial site close to the Poorhouse buildings as a tribute. For a long time, she was thought to be the last Lenape Indian in this area." He gestured to Lee, who joined him beside the fire. "It is now known, however, that there are hundreds of Lenape descendants still here."

Check It Out!

"If there are all these descendants, why do all the plaques say Indian Hannah was the last Lenape here?" Danni whispered to Wyatt.

"Shhh!" said Wyatt.

"The ancient Lenape Prophecy of the Fourth Crow is the story of my people," Lee began. "The First Crow represents the Lenape before contact with the Europeans. We lived in harmony with the Creator and all living things. We hunted, fished, and planted in peace, living according to the seasons and taking no more than we needed. The Second Crow represents the Lenape struggle to maintain their lifestyle during colonization. As a result of the European's clearing of lands and building of mills, our hunting and fishing grounds were destroyed. Many of our people died from diseases brought from Europe and many were murdered by hostile colonists. The Third Crow was a time of madness: the remainder of my people moved west or went into hiding, living like the white man on the outside, but keeping the Lenape way of life alive secretively." Lee paused. "Now, we are in the time of the Fourth Crow, when we no longer need to hide our customs and heritage, but can join all people, in harmony with Creator again, and become caretakers together on this amazing planet. *Welankuntewakan*," she added, bowing her head deeply, as she wished peace to all of Earth's creatures.

The audience responded in unison, "*Welankuntewakan*."

Try It!

A drum beat rhythmically as the children were then instructed, as caretakers, to proceed with the "draping of the gifts for the animals, our brothers and sisters." Wyatt and Danni both reached into the bag of strung fruit and popcorn.

"Yuck," Danni cried as her finger probed a squishy piece of orange.

They walked toward the pine trees behind the fire. The stars sparkled magically. In the eastern sky, Orion the Hunter and Taurus the Bull ascended. Directly overhead, the Milky Way left a ghostly trail across the sky. There was no sign yet of the full moon.

Check It Out!

Danni shivered. Separated from the warmth of the bonfire, the chilly air penetrated her layers of clothing. "I'm going back to the fire," she said, leaving Wyatt with the garlands dangling in his mittened hands. "And hurry up, the potluck starts next."

Wyatt stretched on his tippy toes in an attempt to reach the upper boughs of the pine tree. A shooting star flashed and then disappeared.

"Let me help you," said a kindly voice behind him.

Wyatt turned to see a woman bundled in a parka, her hood hiding much of her face. She took the garland from Wyatt and placed one end a good two feet higher than Wyatt could have reached. She draped it around several boughs before handing it back to him. "This way the birds will get some of the goodies, too," she smiled.

Something about her seemed familiar. As he hung the last of the garlands, he noticed her standing still, looking upward.

"The sky is clear," she said. "*Mushhakot.*"

~ ~ ~ ~

While the adults chatted about the evening's festivities, Wyatt remained quiet during the ride home. Slumped against his shoulder, Danni slept soundly. Wyatt stared out the window at the landscape lit by the rising full moon. The creek sparkled, moonlight flickering off its rolling waters.

Dr. Flo slowed her car in front of the Nystrom car and flashed her lights several times, indicating the first landmark to the left of Brandywine Drive. Luke slowed his car, but didn't stop along the curvy, narrow road. Wyatt saw the sign signifying the Lenape burial grounds that Dr. Flo had been discussing with the Lenape woman.

IN THE WOODED KNOLL ABOVE,
SLEEPING THEIR LAST SLEEP,
REST THE INDIAN OWNERS OF THESE LANDS
BEFORE THE WHITE MEN CAME.

Just up the hill, the ancestors of both Lee and Indian Hannah were buried.

Map It!

The car trailed Dr. Flo's when she flashed her lights again, this time on the creek side of the road. Wyatt got a quick look at another historical marker. INDIAN ROCK was all he could read as they slowly drove past. Indian Rock, Lee had told them, marked one of the last of the Lenape's lands created by an agreement made with William Penn. The Lenape were granted ownership, a concept they didn't really understand, of a two-mile-wide strip bordering the West Branch of the Brandywine from this point to the creek's source in Honey Brook. But it wasn't even three generations of Lenape that made use of the agreement before disease and hostility with the colonists drove them away.

Map It!

Wyatt closed his eyes, visions in his mind of a people, without cars, burdened with their sundries and heavy hearts, leaving this spot along the Brandywine. An image of the woman who helped him drape his garland popped up as he drifted to sleep. She pushed back her hood and Wyatt recognized the gray-braided Indian woman he first saw on a horse, surrounded by dogs and pigs. "*Wanishi,*" the woman in the dream said, bowing her head.

Try It!

Field Trip!

Chapter 18

CHRISTMAS DAY

Christmas morning broke bright, sunny and cold. The sun streamed into Wyatt's room, waking him. He looked at the clock, remembering what day it was, and couldn't believe that he had slept until nine o'clock on Christmas morning. He bolted out of bed and ran to his parents' bedroom. Benny followed behind.

"Merry Christmas," he shouted. He dove into their bed.

"Already?" his father moaned. "It's still dark."

Wyatt yanked the pillow his father held over his head. "No it's not," he said as he and his dad enacted their Christmas-morning ritual. Meg rolled out of bed.

"Let me just run Benny outside quickly," she said. "And then let's have a Merry Christmas!"

By mid-morning, the Nystrom living room had become a chaotic clutter of strewn packages, ribbons, and wrapping paper. Benny lay across a bed of tissue paper. A red bow stuck cockeyed to his head. He chewed on a rawhide as Wyatt leaned against him and fiddled with his new iPod.

"I almost forgot," Meg said. She left the room and returned with an envelope in her hand. "This is from Dr. Flo." She handed it to Wyatt.

Wyatt tore the envelope open and pulled out a letter.

"Dearest Wyatt: I thought and thought about a proper Christmas gift to encourage you to solve the mystery of the woodblock that you and Benny found last summer. I feel rather remiss in not aiding you more than I have so far. Thus, my Christmas gift to you is that of my services, to take you where you need when we are both available: after school when I don't have class or office hours; on weekends; or on days we both have off from school. I will chauffeur you where you feel you need to go, offer my expertise where it is useful, and offer my resources at the University, if appropriate. I give you the gift of my time and my assistance in the pursuit of your hunt along the sweet waters of the Brandywine. Merry Christmas. Fondly, Dr. Flo."

"Wow," said Luke. "That's a generous offer."

"I've got to call Danni." Wyatt jumped up and ran out of the room.

~ ~ ~ ~ ~

The drive to Meg's mother's home in Pennsylvania always seemed long and uneventful to Wyatt, but with his new understanding of the landscape as a watershed system, it held new interest. Each hill they crossed signified a watershed division; each shopping center or development, a barrier to open space; each road, a channel for runoff. He noticed storm drains that he now knew directed runoff directly into the creek. Parks he used to value for their swing sets, now represented open space where precipitation could infiltrate the ground and replenish the groundwater. Train tracks, no longer just an annoying delay when the bells tolled and the lights flashed for an oncoming train, symbolized freight moving from community to community, allowing further expansion and development.

By late morning, the sky had clouded over. Everything looked dreary by the time they reached Modena. The paper mill was deserted, the train tracks were bare, and a hollowness pervaded the borough like a ghost town. The only sound to break the quiet was the car rumbling across the bridge over the West Branch of the Brandywine. Luke parked the car alongside Grandma Burke's home.

Map It!

"Oh, you're here, you're here," Grandma Burke shouted, bursting from her door. She was so full of cheer that even the sky seemed to brighten for a moment. Her arms were extended in greeting as she stepped down from her front porch. "Robert, they're here. Come on out!"

Uncle Robert smiled as he pushed the door open with his foot. An overnight bag dangled from each of his arms and he clutched a loaded bag to his chest. "Thank goodness you're here," he called. "She's been waiting all morning for you guys. Oh, and Merry Christmas, too!"

~ ~ ~ ~ ~

Grandma Burke sat in the front seat beside Luke. Wyatt sat in the back seat, squeezed between his mother and Uncle Robert. Grandma Burke chatted as they drove. "It's a shame about Alex and Jerry's," she sighed as they passed the closed-down tavern. "It was a great place in its hey-day." Luke always took the scenic route with Grandma Burke in the car so she could reminisce. They passed her alma mater, East Fallowfield Elementary, before reaching the hamlet of Ercildoun and heading south toward the former King Ranch.

Map It!

"Remember your friend Emily, Meg? It turns out that the house she grew up in was part of the Underground Railroad." Grandma Burke pointed to an old white house along Route 82. "That one there." Images of quilts with secret symbols and slaves hiding in wagons popped into Wyatt's head. "It was once owned by James Fulton, a founder of this little village. His son, James Fulton, Jr. was an important abolitionist in Pennsylvania."

Meg thought for a minute. "I remember a hidden room in that house where we used to lock Emily's little brother. We'd moan and scratch at the door until he cried."

"Mom," Wyatt reprimanded.

"We were so mean. Wyatt, never do things like that, OK? Do what I say, not what I did."

"You can always be a bad example," Uncle Robert joked.

Grandma Burke continued. "This was an important area for the Underground Railroad. Several of the houses in this village hid slaves. There were a lot of Quakers that lived here and they didn't believe in slavery. After the Civil War, some of these homes were owned by former slaves, including Samuel Ruth whose son William was a great inventor. And look there," she said. She pointed to a blue and yellow historical marker on the side of the road.

IDA ELLA RUTH JONES
THE GRANDMA MOSES OF CHESTER COUNTY

Check It Out!

"Who's Grandma Moses?" Wyatt asked.

"Ah, she was a wonderful folk artist who started painting at my age," Grandma Burke answered. "Imagine that! See, there's hope for me yet."

The family was quiet for a few minutes as they drove through the picturesque landscape. For several miles, rolling green hills bordered both sides of the road. Occasionally, a house appeared, as well as a barn and a post-and-rail fence. When they passed a huge red barn nestled at the base of a hill, Uncle Robert looked at Meg and winked. "Here it comes," he mouthed.

Grandma Burke turned around. "Have I told you about the cowboys in this valley?" she asked.

Meg laughed. "Yes, but I don't think Wyatt has heard it enough."

Grandma Burke looked at Meg. "There's nothing wrong with nostalgia," she said. She looked directly at Wyatt. "About a hundred years ago, this area was owned by the du Ponts." As soon as Wyatt heard the name du Pont, he perked up. He had heard bits and pieces of Grandma's stories about growing up in this area, but never paid much attention. Since starting to investigate his mystery, though, he had a new appreciation for Grandma's insights.

"Is that the same du Pont as the ones who started Winterthur and Longwood?" he asked.

"Oh there are hundreds of descendants, but yes, they are all related. Anyway, the du Ponts purchased about ten thousand acres here for a reservoir. They planned to build a dam on Buck Run and flood this entire area to supply water for Wilmington. Can you imagine this entire area under water if they had built a dam?"

"Why didn't they?" Wyatt asked.

"As it turned out," Grandma continued, "another reservoir was built on the Red Clay Creek, so there was no need to build one here. Thank goodness," she added. "Or I never would have met your grandfather."

Luke suddenly slowed the car down. "Check it out," he said. He pointed to the side of the road. A pack of hounds barked and circled around a group of men and women on horses. The riders wore black helmets, knickers, leather boots, and red coats. Their breath was foggy in the cold air.

"The Christmas Day fox hunt," Grandma explained. "If this was all under water, this certainly wouldn't be taking place."

"Do they really hunt foxes? And kill them?" Wyatt asked.

Uncle Robert answered. "What they do is go to a den and scare up a fox who then runs for his life. The dogs chase the fox and the riders follow the dogs. If the dogs corner the fox, the riders call the hounds off so the fox doesn't get hurt."

"Theoretically," Meg added. "Sometimes the fox gets hurt and sometimes it gets killed. If nothing else, it must get traumatized."

"They'd be more traumatized if they had nowhere to live because this was a lake," Uncle Robert said. "Fox hunting helps to keep the countryside open. And the hunt is for fun."

"Not for the fox," Wyatt said under his breath. Meg smiled at her son and gave him a thumbs up.

Check It Out!

"Where was I?" said Grandma. "Oh, right, about King Ranch. So, since a reservoir wasn't needed after all, the du Ponts sold the land to the Texas-based King Ranch. The rancher wanted to raise cattle up here for the northern markets because shipping them from Texas caused the cattle to lose too much weight." She looked directly at Wyatt. "King Ranch is how I ended up meeting your grandpa."

Map It!

Grandma Burke told her story for the umpteenth time, but, for Wyatt, it seemed like the first time. She recalled the days when all ten thousand acres were spotted with cattle, barns, and cowboys. As a teenager, she and her girlfriends visited a part of the ranch later known as the Laurels, an area lush with oaks, beeches, tulip poplars, and hickory trees. On hot summer days, they swam in Buck and Doe Runs, explored the old Indian trails, and climbed the limestone boulders. Sometimes they'd explore the ruins of the gristmill or the steel rolling mill. One day, she wandered from her friends while they swam in the creek. Wanting to explore more, she climbed deep into the woods where she found a cave, ventured in, sat quietly for a while, and then fell asleep.

"It's been over fifty years and I can still remember the dream I had while I was in that cave," Grandma Burke said. "I wanted to etch it in my mind, so I repeated the details over and over. I didn't know what it meant, but I knew it meant something."

The sound of cows lowing in the distance gave way to a scuffling. Horses neighed in protest as a small, scruffy man led them to the mouth of a cave. A young Grandma Burke scurried behind some rocks to hide. The man hitched the horses inside the cave. She watched as he lit a small fire and sang something over and over, obviously pleased with himself.

"Today I bake, tomorrow I brew
Then I sell a stolen horse or two,
To the British or Americans, I don't care
McCorkle is the name and this is my lair."

"I will be able to repeat that song until the day I die," Grandma added.

"What happened then, Grandma?"

"This McCorkle man saw me watching and got furious. He shook his fist at me and gritted his teeth and said, 'Oh, you're not going to ruin this for me, girlie!' I screamed and woke myself up. A few minutes later, a pair of cowboys appeared at the mouth of the cave. My friends hadn't been able to find me and recruited these two nice cowboys to help look. When they heard my scream, they raced to the cave. They found me exactly where I had fallen asleep: no fire, no scruffy man, no stolen horses. Just a silly girl having a bad dream. Your grandpa said he fell in love with me on the spot."

"My hero," Meg batted her eyelashes and patted her heart.

"Go ahead, laugh," Grandma Burke said. "But if it weren't for that hero, you wouldn't even be here!"

~ ~ ~ ~ ~

After Christmas dinner, Wyatt approached his grandmother, who sat by the fireplace. Off in the kitchen, Meg scrubbed pots and Robert sat on a stool, tossing leftover bits of turkey to Benny, who caught each piece with expertise.

"Grandma, can I ask you something?" he said.

"Of course you can, sweetheart."

"That dream you had, about that guy in the cave. Did you ever find out what it meant?"

Grandma Burke stared into the fire. "Yes, I did, Wyatt. Soon afterward, I learned that the rocks I had been exploring that day were called McCorkle's Rocks. They were named for a horse thief from the Revolutionary War. He stole horses from the neighboring farms and hid up in the rocks before selling them to either side in the war. *Today I bake, tomorrow I brew, Then I sell a stolen horse or two, To the British or Americans, I don't care; McCorkle is the name and this is my lair.* My dream was pretty accurate."

"Do you think he was a ghost?"

"Do I think who was a ghost?"

"That guy in your dream, McCorkle."

"I never thought of him as a ghost," Grandma Burke said. "But more of a vision . . . and a matchmaker."

"Can that kind of thing, you know, having visions . . ." Wyatt hesitated. ". . . be inherited?"

"Why I don't know, Wyatt. Why do you ask?"

"No reason, really," Wyatt lied, thinking of his own visions or ghosts or dreams or whatever the heck they were of Francis du Pont, Indian Hannah, and Horatio Myrick. "It's just so cool that a dream can actually tell you something that's real."

"It is possible," Grandma Burke said, "that subconsciously I already knew about McCorkle, and my dream was taking the subconscious and making it conscious. But that's so much less interesting than the notion that I have secret powers of vision." She winked.

"Can I show you something?" Wyatt said, ready to share with his grandma the tennis ball, the mysterious woodblock and the water-quality data they had collected so far.

~ ~ ~ ~ ~

"What did we do before computers?" Grandma Burke said, as Wyatt booted up. After carefully viewing the woodblock and reading the poem, Grandma Burke had suggested they do some research on the tennis ball.

Wyatt typed into the search bar, "tennis ball USLTA made in USA."

"OK, so the woodblock must have been put into the ball sometime after 1874," Wyatt surmised, after reading that tennis in America began that year.

"Let's try looking up the history of tennis balls in America," Grandma Burke suggested, "since there are a lot of years from 1874 to now."

They searched various websites related to the history of tennis, but didn't get anywhere. Grandma Burke had an idea. "What if we try to buy a tennis ball just like this one," she suggested. Again they searched, but came up with nothing obvious.

"But look." Wyatt pointed to a tagline under one of the hits. "In 1948, the first black person to play in a USTA (known then as the USLTA) . . ." He stopped, recalling that he had noticed that some balls had USTA stamped on them, but his ball had USLTA on it. They searched again until they found a vintage tennis ball site that offered some clues. "Earlier than 1920, tennis in America was run by the United States National Lawn Tennis Association or USNLTA which was changed to the USLTA or United States Lawn Tennis Association from around 1920 to 1975. Later than 1975, tennis was run by the USTA, dropping the 'lawn.' Also around this time, most balls were manufactured in the color yellow to show up better on television sets."

Check It Out!

Wyatt looked at his tennis ball with its blotches of gray and green. Clearly, it had been white at some point. "Well, now we know that the ball was from sometime between 1920 and 1975," he said, disappointed at the broad time frame that offered such little insight.

"Ah," said Grandma Burke. "1920 to 1975: The mystery could have been started in the Roaring Twenties. Or during the Great Depression. Or take your pick of wars: World War II, the Cold War, the Vietnam War. Or maybe it started in the groovy sixties." Grandma Burke made it seem like each era held a world of possibility. "That's when Alex and Jerry's started operating. We spent a lot of time in there talking about the war and the hippies. And talk about crummy water quality. The Brandywine was not good. Do you know that the wastewater from the tavern went directly into the creek without being treated at all! What a mess." She paused before speaking again. "I think you're making some progress with your mystery," she said, holding her right hand up in the symbolic "V" sign for victory. "Peace out, baby. You'll get there."

Wyatt looked at her and then held up his hand in a peace sign. "*Welunkuntewakan*," he said.

Field Trip!

January

Chapter 19

NATURE AT NIGHT

Going back to school after the holiday break was never fun, but the early January field trip for Danni's class made the return from vacation a little less painful. In place of having two days of school in the middle of the month, they would have two evenings of field trips to Paradise Farm Camps to explore the night. There would be owl prowls and astronomy, campfires and hot chocolate. The dead of winter wouldn't be so dead after all.

Map It!

Danni placed her gear in her backpack carefully. It would be freezing outdoors, but the bus ride would be warm, so she knew to pack extra clothes to add or shed layers as needed. The small hand warmers she got as stocking stuffers fit neatly into a side pocket.

To check the effectiveness of her flashlight, she closed her bedroom door and switched off her lamp. With new batteries, the flashlight beam was bright. She turned it off and leaned toward her bedroom mirror. Directing the flashlight toward her eye, she flicked it on. She watched the pupils of her eyes instantly shrink. "The iris is a muscle," her teacher, Mrs. Bibbo, had told them about eye color, "and responds to light. Bright light causes the muscle to expand, shrinking the pupil to keep light out. Darkness causes the iris to contract, enlarging the pupil to let more light in." Fascinated by the involuntary response of her hazel-colored irises, Danni repeated the experiment several times. Then she covered her flashlight with red cellophane, as instructed, and repacked it. She checked the list one last time, satisfied that she was ready to explore the mysteries of the night in comfort.

Try It!

~ ~ ~ ~

Several buses and some large vans were already parked at Paradise Farm Camps when Danni's bus pulled into the parking lot. The Warner Elementary kids filed out of their bus and, like ducklings, followed Mrs. Bibbo along the gravel road. With darkness settling around them, there were no stragglers.

~ ~ ~ ~

"Welcome to all," announced the camp director. Jim was a large man with a booming voice, a friendly smile and a twinkle in his eye. He motioned for Danni's class to stand up. "Let's give it up for Warner Elementary," he bellowed as the kids from the other schools clapped. One by one, he introduced the students from each of the participating school groups: Mary C. Howse Elementary, Church Farm School, Collegium Charter, and Exton Elementary. About one hundred kids filled the open-air gymnasium.

Map It!

In order to get the kids from the various schools to bond, Jim divided them into four groups. Danni was designated a Marshmallow, while others were deemed Chocolate Bars, Graham Crackers, or Campfires. Each group then gathered around one of four colorful parachutes to play games to "break the ice" and get warm.

The gymnasium echoed with shouts and screams as parachutes lifted and kids ran under them, trying to avoid capture by the staff. After a few rounds of parachute games, Danni had made several new friends from various schools.

The staff at Paradise Farm Camps was friendly, enthusiastic, and fun, and Danni looked forward to the two nights of activities with her group. Before going outside, the Marshmallows met in the assembly room for "Owl Prowlers." The students were directed to dissect owl pellets or, as the instructor described, *owl vomit*. Though her new friends wouldn't touch the pellet, Danni dissected it fearlessly, separating the bones, fur, and skull of a regurgitated rodent. Afterward, they mimicked owl calls. "*Who cooks for you? Who cooks for you all?*" they called in imitation of the barred owl.

Try It!

"Neigh like a horse," the instructor said, as they attempted to impersonate screech owls. She then performed a ghostlike trill.

"I've heard that," a girl next to Danni said. "In my backyard. I was spooked by the sound."

The instructor addressed her. "That's another call of the screech owl. It does sound spooky if you don't know what it is."

Danni pictured the old-fashioned boy from Brandywine Creek State Park and the elf-like man from the BVA. *What's spooky,* she thought, *are the things that appear out of nowhere and don't make any sound at all.*

Escorted by a new instructor, they headed outdoors to an open field to look and listen for owls. They fell into a silent line behind Holly, the bird expert, and walked along the perimeter of the field slowly. Finding no sign of owls, they headed into the woods.

Danni yawned. The cold air, late night, and constant activity made her suddenly sleepy. She fell to the back of the line and linked arms with Lindsay, a girl from Exton Elementary, as the Marshmallows entered the woods. Old, dried leaves crunched underfoot. Sticks snapped. Otherwise, the night was still and quiet. The group walked deeper into the woods. Barely noticeable through the thick trees, snow began to fall silently.

"*Whoo, whoo whoo whoo whoo, whoo, whoo, whoo.*"

Danni suddenly felt alert, though everyone had gone perfectly still.

Seconds slipped by. From the opposite direction in the woods, a higher pitched call sounded. "*Whoo, whoo whoo whoo whoo, whoo, whoo, whoo.*"

Moments passed. The deeper pitch sounded again, closer this time. "*Whoo, whoo whoo whoo whoo, whoo, whoo, whoo.*" The group waited for the female Great Horned Owl to return the call, letting her potential mate know where she was.

"*Whoo, whoo whoo whoo whoo, whoo, whoo, whoo,*" she duetted in response.

Silently, a large shape glided overhead and landed on the limb of a dead tree. Holly snapped her flashlight onto a yellow-eyed, two-foot-tall owl staring directly down at the group. Some of the kids gasped at the sight. One of Danni's friends ducked behind her. "Do they eat small people?" she asked. Danni pictured a giant owl pellet housing a small human skull. She shivered and inched closer to her friend.

The owl, uninterested in the humans, lifted his wings and flew off soundlessly. Holly looked at the group. "Lucky kids," she said. Danni wondered if she meant that they were lucky for seeing the owl or that they had brought the luck.

Check It Out!

~ ~ ~ ~ ~

Overnight, four inches of powdery snow fell, but the roads were clear by the late afternoon. As Danni's class walked from the bus to the gymnasium, the snow squeaked underfoot. For the second night, Danni's group would have astronomy, followed by the grand finale: the campfire, complete with hot cocoa and stories.

The Marshmallows made constellation flashcards, played night sky Bingo, and created constellation viewers using toilet paper tubes. They made and ate S'mores.

"Can we have seconds?" Danni's friend Lindsay asked. She told Danni she had never before had the chocolate bar, toasted marshmallow, and graham cracker sandwich.

"You don't say 'can we have seconds,'" Danni said. "What you say is, *I want s'more.* Get it? *Some more!*"

The group sat in a circle to see demonstrations of how the earth moves, causing night, day, and the seasons. They listened to Native American myths about the Milky Way, the Pleiades, and the origin of the moon. Then they ventured outdoors to observe the night sky.

Try It!

Instructor Roger handed out extra pieces of red cellophane for the students who hadn't already covered their flashlights.

"Why do we need the red cellophane?" one of the boys asked.

"It takes about twenty minutes for the pupils in your eyes to fully dilate to adjust to the dark," Roger answered softly, preparing the kids for the quiet hike. "If we shine a regular flashlight near you, your pupils will react to the light and instantly shrink. To see in the dark, it would take another twenty minutes for your eyes to readjust. The red cellophane keeps your pupils from reacting so strongly to the flashlight."

Roger pointed out stars and constellations, using his cellophane-covered flashlight: the brightest star in the sky, Sirius; Orion's belt in the Great Hunter; Cassiopeia along the Milky Way. "Can anyone find the North Star?" he asked.

The kids pointed at different stars, but none of them could locate it.

"It's rather unimpressive, but it's very important," Roger said. He showed the kids how to find it by first locating the Big Dipper and then following the pointer stars of the Big Dipper's cup. "You can look at that star at any time on any night and it will always be in the same spot in the sky. All of the other stars seem to revolve around it. That's why it was used for navigation. Throughout history, people could tell where north is by finding that star. Celestial navigation."

Check It Out!

The term sounded familiar to Danni when she realized that *celestial navigation* was what Mason and Dixon used to figure out the boundary between the free northern states and the slave southern states. And the North Star was used by escaping slaves on the Underground Railroad to direct them north to freedom.

In the adjacent field, telescopes were set up. Danni and her friends took turns looking at Jupiter, the largest planet in the solar system. It had several little moons crossing over it.

"They're so cute," Lindsay said. "Like little baby Jupiters."

"You may think that they're cute," Roger said, "but even the planets played an important role for earlier cultures. Planets were *wanderers* compared to the stars, because they moved against the backdrop of the stars and constellations. The Lenape Indians considered the planets Elder Brothers, supplying night light. Remember, electricity wasn't invented until the eighteen hundreds, and the Lenape didn't have flashlights."

Before letting the students look through the second telescope, Roger had them look at the Pleiades directly. They observed a group of about seven stars. "The seven sisters," he said. Through the telescope, the seven stars looked like a miniature Big Dipper with scores of smaller stars clustered around them.

The Andromeda Galaxy was the focus of the third telescope. It looked like a small, fuzzy Frisbee. "It's the most distant object a person can see, almost three million light years away," Roger said. "What that means is that the light we see now left that galaxy almost three million years ago, and it's taken all that time to reach us. What we are seeing is *history*."

Check It Out!

~ ~ ~ ~ ~

Danni made her way to the seat saved next to Lindsay, carefully carrying her cup of hot cocoa, topped with marshmallows. The campfire warmed her face, though her toes felt stiff and achy from the cold. She sipped her drink and felt the hot liquid run down her throat, warming her upper body. "I should just pour this on my feet," she said to Lindsay.

The camp director signaled the kids to be quiet. "I hear that some of you were lucky enough to see a Great Horned Owl tonight. Most of you know that the owl has long been a symbol of wisdom. The Lenape believed that if you dreamt of an owl, it would become your guardian. Others believed the owl to be a power of prophecy, and helpful in a person's pursuit of the truth. I wonder if any of you were in need of a sign from the owl." He paused. Danni thought about the woodblock mystery. Goosebumps popped up all over her body.

Jim continued. "The world is full of mysteries: the mysteries of space," he said, pointing to the dark sky overhead. "The mysteries of nature. The mysteries held deep within the earth. Mysteries of people and places and history are all around us." Jim told the story of a woman who, over one hundred years ago, brought the first group of children from Philadelphia to Paradise Farm Camps. "Here we are today, together, learning about the world's mysteries because of one woman's vision, generosity and inventiveness. Amazing things can happen when someone has curiosity and takes initiative."

Danni took Jim's words to heart. Maybe the owl appeared for her pursuit of the truth. "Jim," she raised her hand. "Can I share a mystery?"

~ ~ ~ ~

By the time they left the campfire, Danni had a list of recruits promising, come spring, to help gather data to solve the woodblock mystery. Each school would conduct water-quality testing on the Brandywine tributaries close to its campus. Even Jim said he would forward any data collected on Valley Creek by students visiting Paradise Farm Camps.

"What a successful field trip," Mrs. Bibbo commented on the bus ride home.

In more ways than one, thought Danni.

Field Trip!

February

Chapter 20

SNOW DAY

The February snowstorm arrived exactly as predicted. The snow began Tuesday night and fell until early Wednesday morning. By the time Wyatt woke up, the sun was shining and he knew school was canceled.

"Wyatt, you're up," said his mom as he trudged into the kitchen. "Dr. Flo called. She wants to know if you'll go cross-country skiing with her on the Struble Trail in Downingtown. Then she wants to have lunch. Her classes have been canceled until the evening, so she's free today, and thought she'd show you more of the Pennsylvania Brandywine."

Wyatt nodded eagerly. Although he loved the laziness of snow days, he usually ended up bored after an hour or two. "I'll give her a call," he said.

~ ~ ~ ~ ~

The drive paralleling the Brandywine was beautiful. The new-fallen snow glistened in the sun and the water sparkled. Dr. Flo pointed out a red-tailed hawk along the route, sitting calmly atop a leafless tree, scanning for prey.

"The Polar Plunge is next week," Dr. Flo said as they passed Brandywine Picnic Park. "Will you do it with me?"

"What is it?"

"It's a fundraiser for the Brandywine Valley Association, the place where you did the Quest hike, remember?"

"You mean the place with the Myrick ghost," Wyatt said.

"Oh, yes, that too," Dr. Flo said, smiling. "The plunge is great fun. There's a bonfire and people come dressed up in all sorts of crazy costumes. And then, everyone jumps into the Brandywine."

"In February?" Wyatt winced.

"That's what makes it a Polar Plunge. So what do you think?"

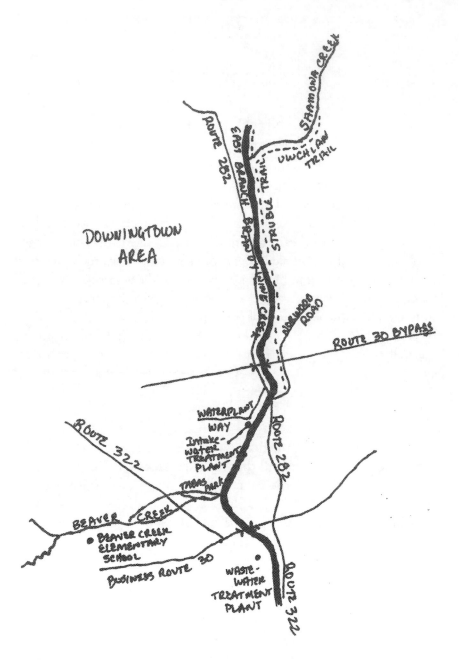

DOWNINGTOWN AREA

"I better check with my mom," Wyatt answered. His mom did, occasionally, come in handy as a stall tactic.

They headed into Downingtown, passing by one of the most inviting playgrounds Wyatt had ever seen. A series of turrets and bridges connecting monkey bars and gazebos reminded him of a wooden fortress. EAST WARD ELEMENTARY SCHOOL read the sign. They continued past an old library.

Map It!

"They say there's an underground tunnel that connected that house," Dr. Flo pointed to an old Federal-style home on their right, "to the library," a quaint stone building on the left. "Now why would there have been a tunnel connecting those buildings?"

Wyatt thought for a moment. "Well, we just passed a Quaker meetinghouse, so there were probably Quakers who lived around here. And Quakers didn't believe in slavery, so maybe the tunnel was used to help the slaves escape."

"I bet you're right; it was probably part of the Underground Railroad," said Dr. Flo.

They approached a bridge crossing over the Brandywine and passed a very old and very small log house. "That is the oldest recorded house in Chester County," she said. "It was built in the early seventeen hundreds, when this road was just a dirt path. If we wanted to cross the river, we'd have to ford it in our horse-drawn wagon because there was no bridge."

Map It!

Wyatt imagined a dirt path surrounded by forest where there was now a paved road bordered by municipal buildings, sidewalks, power lines, and antique shops. He pictured children spilling out of the little log cabin, heading to the creek to haul water for bathing and cooking. Or maybe they'd be splitting logs to keep the stove stoked through the cold winter. He'd wave to the kids from the bench of his Conestoga wagon. Then he'd splash through the shallow stretch of the Brandywine on his way to Lancaster. Or, in the eighteen hundreds, a century and a half later, he might be driving his hay wagon over an actual bridge toward the library, transporting slaves through the night as they hid under the hay.

Check It Out!

~ ~ ~ ~

Wyatt clipped his skis into their bindings and followed Dr. Flo as she set out on the Struble Trail in Kardon Park. The pond shimmered in the bright sun as ducks and geese of various colors and sizes paddled around. He maneuvered into Dr. Flo's tracks awkwardly and began the slide, kick, slide, kick rhythm of their cross-country trek. Bare sycamores, with their camouflagelike bark, bordered the stream. Except for a set of dog-paw prints and some previously laid ski tracks, the trail appeared to be deserted.

"Do you know which branch of the Brandywine we're on, Wyatt?" Dr. Flo asked.

"The East Branch," he answered confidently. After Danni told him about all of the schools from her field trip willing to help out with data collection, he had checked a map. The schools were all located near tributaries that flowed into the East Branch close to Downingtown.

"Righto." Dr. Flo began singing softly. "*When you're alone and life is treating you lonely, you can always go, to Downingtown.*" She stopped singing. "Do you know what this town was called before it was Downingtown?"

Wyatt shook his head.

"I'll give you clues along the way," Dr. Flo said. "Your first clue is the old paper plant that is now a restaurant right in town." She pointed behind her.

"Paperville?"

"Don't be so impulsive. Wait for a few more clues."

The trail passed under the highway and their voices echoed in the graffitied underbelly of the bridge. JS & WN 2GETHER 4EVER. Wyatt thought about Jennie, the girl from school that supposedly liked him more than as just a friend. "Your friend Jennie," Dr. Flo said as though she had been reading his mind. "What's her last name, *Wyatt Nystrom?*" Wyatt blushed, quickly looking away from the graffiti. "Shaw," he replied quietly.

They continued to ski along the creek, passing a broken-down dam, a rope swing, and a campground abandoned for the winter and blanketed with snow. "Clue," Dr. Flo said, pointing at the dam.

"Dam town?" Wyatt said.

"No, and watch your mouth, young man," Dr. Flo joked.

They passed the ruins of the Mary Ann Forge, which Dr. Flo designated as another clue. Skiing quietly along the Brandywine, Wyatt listened to the juncos and chickadees as they searched for food. The snow along the trail sparkled in the sun.

Dr. Flo then led Wyatt up a short detour on the Uwchlan Trail and over an iron pedestrian bridge. Shamona Creek, a tributary to the East Branch of the Brandywine, gurgled beneath them as the water flowed around icy rocks. Set back in the shade among snowy shrubs and trees stood the remains of an old gristmill. "Clues, clues, everywhere!" Dr. Flo said.

Map It!

"Milltown?" Wyatt guessed.

"That's my boy! You got it. Milltown was home to a number of mills until very recently. But the name changed in the late seventeen hundreds to honor the Downing family, who not only owned a complex of mills along the Brandywine, but lived in that log house we saw earlier." She changed the subject. "Now we can go up this trail along Shamona Creek, or we can go back on the Struble Trail along the Brandywine as far as the next old paper mill. This trail's got a bit of a hill, but there are some great historic markers related to this gristmill. Or," Dr. Flo said, winking, "we can turn around now and head back for some food. Not to influence you, but food has got my vote."

Check It Out!

"Food sounds good to me," Wyatt answered. *And I think I've seen enough mill ruins for a while.* As he turned his skis around, he saw two crows fly past and land on the old mill.

~ ~ ~ ~ ~

The Coffee Cup Restaurant in Downingtown was crowded, so Dr. Flo and Wyatt sat at the counter. A friendly, young waitress approached. Her nametag read, "Kat."

"What can I get for you sweeties?" Kat asked.

Wyatt thought she seemed kind of young to call them *sweeties*, but it didn't seem out-of-place here. On Dr. Flo's recommendation, he ordered a *mess* of eggs with potatoes and onions. He then spun around on his stool and headed off to the men's room.

"Well, hello young man," said an elderly man, as Wyatt returned to his stool. Wyatt smiled tentatively before Dr. Flo introduced him.

"Wyatt, this is my dear friend Claire, or Mr. Pearson. We went to college together."

"A long, long time ago," Claire added, smiling. His forehead crinkled and Wyatt noticed that one of his eyes was cloudy.

"We were just talking about old times," Dr. Flo added.

"Back in seventy-two, Hurricane Agnes. I'll never forget it. My father's car was filled with water right outside this store. And in ninety-nine, it was Floyd, but that didn't do as much damage on account of that dam up at Marsh Creek."

Check It Out!

Wyatt nodded agreeably, though he had no idea what cloudy-eyed Claire was talking about. When the waitress set down his *mess,* he dug in, eager to focus on something other than floods. *After all, dams were for mills, weren't they? What did dams have to do with floods?*

"Well, I better be gettin' on now, ole Flo. It was great seeing you again." Claire gave Dr. Flo a kiss on the cheek. "So long, kid."

Wyatt waved goodbye with his left hand because he was busy writing on his napkin. CP & FW 2GETHER 4EVER.

"Unlike your friend Jennie," Dr. Flo said, "Mr. Pearson is a married man."

~ ~ ~ ~ ~

Dr. Flo pulled her Prius into a space close to the main entrance of Beaver Creek Elementary School. "There's someone I want you to meet," she said. Though the parking lot was plowed, there were only a few cars.

The custodian greeted them as they entered the school. On the mat, they stomped bits of snow from their shoes so they wouldn't slip on the polished floor.

"Knock, knock," Dr. Flo said, lightly tapping on the open door labeled PRINCIPAL LAWLESS.

What kind of name is that for a principal? thought Wyatt. *It sounds like there are no rules here.*

"Come in," Dr. Lawless said. "How are you, Flo?" She stepped from behind her desk and embraced Dr. Flo. Wyatt looked at the desk cluttered with papers, plastic cups, measuring containers, and a plastic bottle of blue-colored water.

"Great," Dr. Flo said. "Look at you, now a principal. How is it going?"

"It's terrific, Flo. I just love it. Though it's a little quiet for me today, being a snow day. And who is this?" She nodded to Wyatt.

"This is my friend Wyatt."

The principal and student greeted each other before Dr. Lawless and Dr. Flo launched into conversation that seemed to Wyatt to encompass everything from family vacations to global warming. When the conversation turned to Dr. Lawless's husband and his annual frog legs dinner, Wyatt decided to go exploring. He wandered down the corridor, peeking into deserted classrooms and checking out the artwork posted along the walls. A banner reading WHAT WE USE WATER FOR hung over first-graders' drawings of children drinking water, washing their hands, cooking pasta, washing the dog, swimming in a pool, and brushing their teeth.

Check It Out!

"Wyatt," called Dr. Flo from the other end of the hallway. "Dr. Lawless needs some help."

Wyatt ran back down the long corridor, figuring that if there was ever a time he could get away with running in the halls, it was now, on a snow day, with a principal named Lawless.

"Dr. Flo told me about your mystery," Dr. Lawless said. "And she tells me you've been recruiting schools to help gather water-quality data.

You know, this school is right next to Beaver Creek, a major tributary to the East Branch of the Brandywine. Could you use data on it?"

Map It!

"Definitely."

"Good. Now, if I'm going to have my students gather data for you, can you do me a favor? I'm checking a proposal for an in-service for my teachers, and I want to know if kids would like some of these activities. Can you help me?"

"Sure."

She pulled a roll of toilet paper off of her desk. "Now, Wyatt, do you mind if I wrap you up in this?"

Wyatt just shrugged.

"OK, let me know when you think I've reached the point that represents how much water is in your body." Starting at Wyatt's feet, she carefully wound the roll around him, shrouding him in toilet paper. When the paper ripped, she reattached it with a piece of tape.

When she reached Wyatt's chest, he spoke. "Now is good."

"About seventy percent. Good job. What do you think of that as a demonstration?"

"Pretty good," he answered. "But maybe use something that doesn't tear so easily, like fabric. And maybe it should be blue for water."

Try It!

"I told you he's smart," said Dr. Flo.

"OK, one more," Dr. Lawless said to Wyatt. She showed Wyatt the bottle of blue water. "If this bottle of colored water represents all the water in the world, how much do you think is potable?"

"*Potable*?" Wyatt asked.

"Yes, meaning it's fresh and clean enough to drink and use."

Wyatt thought about the first-graders' drawings in the hallway. He poured out a cupful.

"OK," Dr. Lawless said, scrutinizing the lesson plan. "Let's see if you're right." She poured the cupful back into the bottle and proceeded to measure out various quantities of water representing salt water, polar ice caps, polluted fresh water, and deep groundwater. Only several drops were leftover. "Of all the water in the world, less than one percent is available to us as fresh water for all of our daily needs. That's amazing."

Try It!

"And alarming," added Dr. Flo.

"If there are about seven billion people on earth, then how can we possibly always have water available when we turn on the tap?" said Dr. Lawless.

"The water cycle," Wyatt said.

Check It Out!

Dr. Lawless looked embarrassed. "Yes, of course," she answered. "I just wanted to make sure that you knew it."

~ ~ ~ ~ ~

Heading homeward, Dr. Flo stopped at a small field along Beaver Creek. TABAS MEMORIAL PARK, read the sign.

Map It!

"See the dam there?" She pointed out the broken concrete structure in the creek and the ivy-covered gristmill ruins beyond it. "The Roger Hunt mill was built close to three hundred years ago. Farmers from the area would come here to get their wheat milled, whiling away the hours in the local tavern, since there wasn't much else to do."

Wyatt pictured the dirt road jammed with horse-drawn Conestoga wagons. Some would be waiting with their wheat, while others would be exiting, their wagons filled with flour. The drivers would either be happy and tipsy from partaking in libations, or grumpy and hung-over from celebrating too much the night before.

Dr. Flo interrupted Wyatt's thoughts. "There's one last thing I want you to see while we're in the area." They took a slight detour up Route 282 and down a small street by the name of Water Plant Way. They parked by the Brandywine.

The creek sparkled, gurgling its way between the snow-covered banks. "Look there." She pointed toward the creek to a large, submerged silver container. "That is where water is taken from the Brandywine and sent to the treatment plant to be processed for Downingtown's potable water. About a mile downstream is Downingtown's wastewater treatment plant."

Map It!

"It's a good thing the wastewater plant is downstream from the drinking-water plant," Wyatt said.

"Yes, it is. But Downingtown's wastewater treatment plant is still upstream from West Chester's and Wilmington's water treatment plants."

Map It!

"Yeah, but I guess I'm glad that I have water at all, after what Dr. Lawless showed us. And at least I have running water at my house," he said, thinking of the historic log house, "even if it has already been used."

Field Trip!

Chapter 21

TAKING THE PLUNGE

Wyatt jumped up and down, trying to warm up. "If you're cold, keep your body moving," Dr. Flo told him. He wasn't sure if it was because of nerves or cold, but he couldn't stop shivering.

"Are you sure you want to do this?" Meg asked her son. "You don't have to, you know."

Wyatt looked at her. "I've gotten this far, Mom. I can't give up now."

A man with a can of spray paint pointed to Wyatt. "Next!"

Wyatt handed his bathrobe to his mother. In just his swim trunks and water shoes in the middle of February, he sure felt exposed. He turned his shoulder to the man who sprayed it with the number thirty-three, the temperature of the water that Wyatt had agreed to plunge into.

Brandywine Picnic Park buzzed with anticipation. A man who was dressed appropriately for a chilly winter morning lit the bonfire and made an announcement for the costume contest. Following that would be the plunge, and Wyatt braced himself for the run into the creek with hundreds of other crazy plungers. His friends, after hearing that an "old lady" had invited him to participate, wouldn't let him off the hook. In return, he made them pledge money in support of his plunge. That "old lady," Dr. Flo, now stood by the fire holding a golden toilet plunger, ready to present to the winner of the costume contest. Among the snowmen, ducks, beauty queens, and cartoon characters, Wyatt favored the pirate team, Team Aaargh.

"Plungers, please line up." The announcement over the loud speakers startled Wyatt, even though he knew it was coming. EMTs, dressed in wetsuits, were already in position at posts along the creek. An ambulance idled nearby, and a group of black-leather-clad motorcyclists pulled up to watch. Across the creek, on the road overlooking the scene, several cars were parked and onlookers watched from behind the guardrail.

Wyatt found Dr. Flo. They approached the starting point and blended into the mass of scantily clad bodies. Checking that it was as cold as he thought, he blew into the air and watched as his breath formed a cloud of condensation.

"At the sound of the gun, you are to enter the creek, cross it, and return," the announcer continued. "Are you ready?" A timer beeped loudly, counting down: five, four, three, two, one. A gun blasted and hordes of plungers jumped from the bank into the frigid water. Shrieks and laughter filled the air as more and more people entered the creek. Those that entered first crossed to the far bank. Dr. Flo grabbed Wyatt by the hand and took the plunge. "We're doing it for the critters!" she yelled as they tried to cross the creek. But the force of the moving water slowed them. Wyatt stumbled on the uneven bottom and tumbled into the water, bringing Dr. Flo down with him. He got up. Downstream, in the crook of a tree, he saw four crows watching.

Dr. Flo heaved herself from the water. She turned toward Wyatt and took his hand again. Pumping their fists in the air, she laughed and yelled, "For the critters!" and they continued across the creek. *Madness*, Wyatt thought.

Field Trip!

Spring

March

Chapter 22

LEGACIES OF PRIDE AND SHAME

Danni lucked out. Delaware had a day off from school, but Pennsylvania didn't. Since both of her parents had to go to work, she was to spend the day with her cousin Quinn, in Coatesville. At first, Danni was upset that she had to accompany her cousin to *her* school, but when she found out the class was going on a field trip, she figured it might not be so bad after all.

Normally, Danni would have stayed with Wyatt, but he was sick. It seemed ironic that he was completely healthy following the Polar Plunge back in February, but now, on the first official day of spring, he had the flu.

Along the road to Rainbow Elementary, bright-yellow forsythias and pearly-pink magnolias were a welcome sight after the long winter. Danni loved Quinn's school immediately. Not only was it cheery and bright, it looked inviting and smelled new.

Map It!

Danni waited in line behind her cousin to board the school bus. Quinn's friends seemed nice, once Danni could keep straight who was who. And the day would pass quickly with a trip to the Lukens National Historic District in Coatesville, followed by a picnic lunch and nature hike at the Saalbach Farm. Danni had never heard of either place, but figured it would be another adventure, not unlike the ones she'd been having all year, since she and Wyatt had started looking into the mystery of the woodblock. She thought about the inscription that had lodged itself in her mind:

~~Gladness~~
Sadness
Badness
Who will help to stop the Madness?
Y
ch44

"Hey, girl, quit spacing out!" Quinn called to her from the bus. Danni looked up to see that she was holding up the rest of the line.

~ ~ ~ ~ ~

The bus turned right onto First Avenue. A colorful globe welcomed them to the historic Lukens district in Coatesville. Above the globe, several flags fluttered in the warm breeze. The bus parked in the lot, and the fifth-graders of Rainbow Elementary clamored out before being divided into tour groups. Behind them were deserted factory buildings with a sign reading, FUTURE SITE OF THE NATIONAL IRON AND STEEL HERITAGE MUSEUM. In contrast, a stately brick, slate-roofed building stood in front of them.

Map It!

The group entered the brick building. An elegantly dressed gentleman descended a sweeping staircase, bowed deeply, and introduced himself. Danni felt like she had stepped back in time to the early eighteen hundreds.

"Do not let me frighten you," he began. "Though I died many years ago."

The hairs on the back of Danni's neck tingled. *Here we go again,* thought Danni.

The man continued. "Allow me to introduce myself. I am Dr. Charles Lukens, proprietor of this amazing industrial complex, once known as Brandywine Iron Works, later known as Lukens Steel. Now we are known to the good people of Coatesville as ArcelorMittal." He led the group past a wall of sepia-colored photos to a portrait of a woman. "My beautiful bride, Rebecca Pennock Lukens, ladies and gentlemen. The mother of my children, the bright and inquisitive daughter of Isaac Pennock, and the first female to run an industrial empire after my untimely death at the age of thirty-nine. Be on the lookout for her; her ghost sometimes becomes restless with memories of the ups and downs of this industrial empire and its community."

I knew there'd be ghosts, Danni thought. She stared at the portrait of Rebecca Lukens. She wore a high-collared lace gown that looked more like a nightgown than a dress. She had a full face, widely-spaced eyes, and dark hair covered with a net. She was the image of any old-fashioned wife or mother, yet this woman was the first female industrial leader in America. She was a big deal.

Danni and Quinn partnered up for a family tree activity. They were to complete a fill-in-the-blank worksheet by getting clues from the pictures, portraits, and displays around the room.

When they completed the assignment, they turned their paper in to Dr. Lukens, who put on his spectacle and examined their answers. Danni stood on her tiptoes to read alongside him.

"Very nice," he commented. "Yes, yes, nice job."

A LITTLE BIT OF LUKENS HISTORY

Check It Out!

1. *Isaac Pennock owned a mill on Buck Run, a few miles south of the town of* **Coatesville**.
2. *In 1794, Pennock and his wife, Martha Webb, had a baby,* **Rebecca Webb Pennock**.
3. *As a girl, Rebecca learned a lot about the family business of running a mill that produced* **iron wheel rims** *and* **barrel hoops**.
4. *When she grew up, Rebecca married a man named* **Dr. Charles Lukens**.
5. *Isaac Pennock bought a sawmill in Coatesville from Moses Coates (for whom Coatesville was named). The mill was on the* **West** *Branch of the Brandywine. Pennock converted the sawmill to an iron mill to make products for the growing railroad and shipping industries.*

6. *Pennock hired Rebecca's husband,* **Dr. Charles Lukens**, *to run the iron mill, Brandywine Iron Works.*
7. *In 1825, Dr. Charles Lukens died at the age of* **39**.
8. **Rebecca Webb Pennock Lukens** *replaced her husband as the head of Brandywine Iron Works, making her the first female in America to run an industrial company.*
9. *Her company, later named* **Lukens Iron Works**, *became a steel mill, specializing in boilerplates used for steam engines for ships and locomotives.*
10. *Rebecca Webb Pennock Lukens raised* **2** *daughters to adulthood.*
11. *Her daughter, Isabella Lukens, married* **Dr. Charles Huston**.
12. *When Rebecca Webb Pennock Lukens died in 1854, at the age of* **60**, *her son-in-law took over the Lukens industry.*
13. *Rebecca is buried in a graveyard in the borough of Ercildoun, just south of* **Coatesville**.

Danni noticed that there were thirteen items on the worksheet, and the last one was about a graveyard. She shivered, suddenly feeling cold.

～ ～ ～ ～ ～

Danni's group crossed the street and entered the elegant Graystone Mansion, built in the late eighteen hundreds by Rebecca Lukens's grandson. The foyer was decorated with diagrams depicting the iron melting and rolling processes. Lining the walls were models of ships and locomotives that used Lukens's iron plates. The group watched a film about the "oldest continuously operating steel mill in America," that made the point that, in the early eighteen hundreds, most transportation was by foot, horseback, or carriage. There were no cars, traffic lights, sirens, and drive-through ATMs. When people started using railroads, the need for iron products grew. Lukens made those products in their iron mill by using waterpower from the West Branch of the Brandywine Creek.

Outside the mansion, sunshine filtered through the leafless sycamores that lined First Avenue. *Camouflage trees*, Danni remembered Dr. Flo calling them. The peeling bark resembled the fabric designed for military use and was now so popular that some of the girls at school wore a version in pink. After strolling under the trees, the group reached the original home of Rebecca and Charles Lukens, a rundown, old building, considering it was called the "Brandywine Mansion."

The tour guide explained something about the building's history and its various inhabitants. One of the owners was Moses Coates, who had run the original sawmill that evolved into the Lukens's iron-to-steel empire.

"It's not a very nice place to live for such important people," Quinn whispered, but Danni was distracted. In the topmost window of the house, a woman peered down. Though partly hidden by a lace curtain, she wore a net over her head and was dressed in what appeared to be a nightgown.

~ ~ ~ ~ ~

Across from the Brandywine Mansion, a man and woman waited for Danni's group. They stood on the porch of an elegant house adorned with curved windows, dormers, and finials. Of course, they were dressed in vintage clothing.

"Now, *these* people had some cash," a boy in the group joked.

The costumed man stood and introduced them as "Isabella Lukens and Dr. Charles Huston, daughter and son-in-law of Rebecca Lukens." He explained that this home, called Terracina, was a gift to them from Rebecca. After she died, Dr. Huston took over the mill operations and became its first official president. At this point, Danni was getting a little tired of olden-days talk and was losing interest until Dr. Huston mentioned something that caught her attention. "The property," he said, "has many horticultural delights. But what we really enjoyed was this open area." He gestured to the expanse of well-manicured lawn and took a tennis ball from his pocket. "This is where we partook of our favorite pastime, lawn tennis."

Danni could see it clearly from where she stood. "USNLTA." She waved her hand impatiently. "Excuse me," she said. "When did you play tennis here?"

"Ah, a tennis player, I presume?" the man replied. "Oh, I'd say I was playing in the eighteen-nineties or so, back in the days when we did indeed play on the lawn. It wasn't until later that courts were created for the game."

Danni took a pen from her back pocket and wrote on her wrist: "USNLTA: check with Wyatt."

~ ~ ~ ~ ~

"I'm starving," Quinn said to Danni as they sat at a picnic table at the Saalbach Farm.

Map It!

The girls unpacked their lunches and looked over the contents before deciding what to eat first. Danni looked at Quinn's brown-tinged apple slices. "It's better than *that*," Quinn said as Danni pulled out her blackened banana. One seam of the peel had split, and a gooey pulp oozed out.

"I'll just stick with the pb and j," Danni said. *When will moms learn not to pack bananas in bag lunches?*

Danni stayed with Quinn and her group for the after-lunch field study. Joanna, the field instructor, distributed buckets of equipment for the kids to carry. She guided them across the road and along a trail on the old farm property. Juncos, finches, and a pair of cardinals flitted along the hedgerows.

Check It Out!

"The bright-red bird is the male, and the more boring-colored one is the female," Joanna explained. "Does anyone know why the female is less colorful in the bird world?"

"So the females don't attract predators," called out a girl named Emerson.

"So the mother can stay camouflaged with her eggs in the nest," called another girl.

"Because boys rock," called out a boy.

"Did you know that," Joanna began, "in addition to being more colorful, it's the *male* songbirds that sing, not the females? It's the same for frogs. Only the *males* sing to attract the females. Of course, with humans, it's the *females* who do all the *looking good*."

Check It Out!

As they headed downhill into the woods, the ground grew increasingly moist. One boy accidentally stepped on a large, leafy plant growing from the mud.

"Yuck! What's that smell? It smells like skunk," he said.

"You got that right," Joanna called out pleasantly as a very unpleasant stench wafted by. "It's skunk cabbage." Danni stepped around the offensive plant right into the thick, black mud. Her shoe held fast, and she stood glued to the spot by suction. Joanna gave her a hand and pulled her out, but the shoe stayed behind.

"Somebody, get my shoe!" she called, hopping on one foot and giggling. The others laughed, too. Quinn managed to pull the mucky shoe out of the mud.

"You're my kind of girl," Joanna said with a smile. "Getting down and dirty like a true outdoors girl. A little shoe in the wetlands isn't going to stop you." Danni thought about the wetlands visit from her art class, where they saw the bald eagles. *Wetlands mean healthy habitat*, she recalled, *not just gross sneakers.*

Joanna led the group to an opening along Rock Run and described the methods they would use to do the water-quality testing. Danni recognized the procedures from what she and Wyatt had done in the fall with Dr. Flo. Joanna added that, although it wouldn't be a precise scientific study, enough information could be gathered to make a general assessment of the creek's water quality.

Several of the kids in the group volunteered to measure the physical aspects of the creek, including depth, width, velocity, turbidity, and temperature. Danni nudged Quinn to volunteer for the chemical tests. "I've actually done these before," Danni told her.

After recording their results, Joanna described the method they would use to collect the macroinvertebrates.

"What's a macroinvertebrate?" Quinn asked.

"Oh, I know, I know," Danni called out. Joanna looked impressed and nodded for her to explain. "It's an animal that's big enough to see, but that doesn't have a backbone."

"Exactly," Joanna said. "They're the little critters we will be looking for under the rocks. Not only do they provide food for the salamanders and frogs and fish, but they indicate how clean the water is."

Joanna distributed cups and paintbrushes to the kids and went over the rules of critter catching, including placing rocks back where they were found. Then she let the kids go hunting.

Shrieks sounded as the students discovered mayflies, stoneflies, planaria, and caddisflies crawling and clinging to the bottoms of rocks. One boy caught a crayfish, but dropped it when it pinched him with its claw.

Danni lifted a large rock from some fast-moving water. She looked at its underside. "Oh, snap," she exclaimed, nearly dropping the rock.

"What?" Quinn asked, making her way across the rocks.

Danni told Quinn to get the paintbrush and cup of water ready. She pointed to the two-inch-long, thorny-sided creature with pincers on its head scurrying across the bottom of the rock. Quinn screamed.

"Get it in the cup," Danni yelled. Quinn brushed the little beast into the cup of water. It thrashed around as Quinn held it at arm's length. "What the heck is it?"

Joanna rushed over to see what the commotion was about. "Beautiful!" she cried. "Let's get him into something bigger." Quinn gladly handed the cup to Joanna, who placed the hellgrammite into a larger container.

"The great white shark of the creek," Joanna said. After gathering the students together to look at the macroinvertebrates they had found, she passed the cup with the hellgrammite. Some of the students wouldn't even touch the cup. "It's actually the larval stage of the dobsonfly. Just like caterpillars metamorphose into butterflies, hellgrammites metamorphose into airborne dobsonflies. They live their *youth* in the water, and spend their *adulthood* on land and in the air. Listen." She stopped talking momentarily. The twittering and chatting of sparrows, wrens, chickadees, and nuthatches, filled the air. "In the larval stages, they are food for fish and salamanders. As adults, they become food for birds."

Check It Out!

The group passed around the cups. The mayflies, stoneflies, caddisflies, craneflies, planaria, worms, snails, water pennies, and hellgrammite indicated good water quality in this tributary to the Brandywine's West Branch. Quinn recorded all of the data on the form Joanna had given her. When the group returned to the bus, she handed it to her teacher. Danni watched as the teacher dropped the paper into a trashcan beside the bus. When no one was looking, she carefully retrieved the valuable data.

~ ~ ~ ~ ~

"How was the day with Quinn?" June Nystrom asked her daughter once they got into the car. Danni was excited to tell her mom about the field trip and how she even got a set of water-quality data for the mystery she and Wyatt were trying to solve. Less exciting was the field trip to the historic Lukens district, but all in all, it was a good day.

"I wasn't aware that you were going to the Lukens site," her mother said. "Your great-great grandfather Isaac worked there nearly a hundred years ago."

"Did he work for that lady? The first lady to run a big company like that?"

"Who? Rebecca Lukens? No, she died long before your great-great grandfather Isaac worked there. And lucky for her that she didn't live long enough to witness what *he* witnessed."

"What do you mean by that?" Danni recalled the ghostlike figure in the window that she had seen earlier that day, and thought of Dr. Charles Lukens's warning: "*Her ghost sometimes becomes restless with memories of the ups and downs of this industrial empire and its community.*"

June told the story that had passed down through the generations of her family. Though she was warned that it was an unpleasant story, Danni insisted.

"I guess it's time that the story be passed on to the next generation," June said, "in hopes that nothing like it ever happens again."

The year was 1911: Young Isaac Jones made his way each weekday to the Lukens steel mill to work on the pouring line. The grandson of slaves, he moved up from North Carolina a few months earlier and quickly found work at the mill, though he could have just as easily gone to the Worth Brothers' steel mill nearby. Like so many black men from the south and immigrants from Europe, he found work in the booming industries of the north, but he didn't exactly feel welcomed by the residents of the mostly white town. Still, he had a job to do, he did it well, and he tried not to cause any trouble.

But trouble began anyway. August 12 was hot and humid. Like most Saturdays, most of the workers were in town, cutting loose at the local taverns. In the corner of a smoky bar, Isaac drank whiskey and swapped stories with a group of black steel workers from the Worth mill as the sun journeyed toward the western horizon.

Hours later, Isaac heard the bad news. One of the Worth workers he'd been drinking with, Zacharia Walker, got looped and, on his way home, shot his pistol at some Polish steelworkers. A security guard, Edgar Rice, heard the shots and went to investigate. He got in a scuffle with Walker. Rice, a white man, was shot and killed. Walker took off, scared, and hid in a tree. Knowing he was in big trouble, he tried to kill himself but missed, shooting himself in the jaw. When he fell from the tree, the authorities took him to the hospital.

But it got worse, lots worse. Isaac heard how it unfolded as the news spread through town, word-of-mouth, like wildfire. Walker was shackled to his hospital bed, where he confessed to the shooting in self-defense. But word on the street was that Walker was a cold-blooded killer. A mob broke into the hospital and dragged the shackled Walker, bedpost and all, out to the edge of town to a clearing. Perspiring from fright as much as from the hot, humid weather, Isaac stayed out of sight with the other black men and kept their distance from the raucous crowd that gathered. Thousands of white men, women, and children traveled to Coatesville by train, wagon, or on foot to witness the spectacle as if it were a football game or horse race.

Isaac heard the screams and could only imagine the gruesome scene. A bonfire had been lit, and Walker was launched into the flames, his leg still shackled to the hospital bedpost. They heard the details later: how Walker had pleaded not to be given a "crooked death"; how he begged for mercy, having shot Rice in self-defense; how he crawled, charred, from the fire and was pushed back into the flames; how he managed to struggle out of the fire two more times before being roped like an animal and dragged into the inferno, until his screaming body finally acquiesced and sizzled into a human roast.

Monday morning, Isaac heard some of his coworkers talking. Turns out, some of the white kids from the mob returned to the scene for souvenirs. "Yeah, my kid came home last night with a roasted toe," a white man boasted.

Check It Out!

Danni shivered, feeling a little sick. "That's so horrible."

Danni's mother squeezed her hand. "No one was ever convicted for the crime of lynching Zacharia Walker," she said. "But it is a tragedy people need to know of."

Now that sounds like madness, thought Danni.

Field Trips!

April

Chapter 23

EARTH DAY BIRTHDAY

Wyatt's father woke the cousins. "Rise and shine, guys! We have a birthday to celebrate today!"

"What?" Wyatt said groggily, sorry that he had stayed up until two in the morning.

"It's Earth Day's birthday," Luke said cheerily.

Although Wyatt and Danni had been looking forward to this day all week, they moved slowly through the morning's routine. *Why do they call them sleepovers?* Wyatt thought. *They should be called awake-overs.*

~ ~ ~ ~ ~

Before going to the Earth Day event, the Nystroms had decided to stop at Maysie's Farm Conservation Center to check out the organic farm. They had discussed joining one and took the opportunity to visit Maysie's since it was just a few miles from the Earth Day birthday celebration.

Map It!

Luke parked between a greenhouse and a large solar panel. In front of them, fields divided by crop rows spread in all directions. The sun poked out from behind cotton balls of cumulous clouds. Danni closed her eyes and turned toward it to absorb the warmth. Something wet touched her hand.

"That's Duke," a white-bearded farmer called from the barn. "He's friendly."

Danni petted the dog, and Wyatt hurled an old tennis ball down the grass swath that divided the vegetable beds. Duke chased the ball, then crouched to monitor its moves.

"He's an Australian shepherd," the farmer said. "He's not a fetcher; he's a herder."

Check It Out!

The cousins chased after Duke, and Wyatt's parents walked over to talk to the farmer. There were no leaves on the trees yet, but the maples had flowered and there were velvety gray buds on the pussy willows. Overhead, a group of turkey vultures circled, catching thermal drafts and kettling upward.

"I forgot to tell you about the tennis ball," Danni said. Wyatt, amused by Duke's herding instincts, threw the ball again. "People on our field trip last month said something about there being lawn tennis before there were tennis courts."

"Didn't I ever tell you?" Wyatt said. He suddenly felt guilty. "Me and Grandma Burke figured out about the tennis ball. USTA is the United States Tennis Association; that's what it's called now. What *our* tennis ball has on it is *USLTA*. That's what it was called before USTA."

"The ball the guy had at Lukens said *USNLTA*,'" Danni said. She had committed the lettering to memory, but had kept forgetting to check it out with Wyatt. "He said that they played on the lawn."

"How old was the guy?"

"I don't know how old he really was, but he was supposed to be from the eighteen hundreds."

"So," Wyatt said. "It turns out that USNLTA was the name used *before* USLTA. It stood for United States National Lawn Tennis Association. USLTA stands for United States Lawn Tennis Association and was used from about 1920 to 1975. USTA is the name they started using in 1975 and still use now."

"I can't believe you never told me," Danni said. "Here I am getting all these different schools to send in information to solve the mystery, and you don't even tell me something that important? You stink!" She punched Wyatt's shoulder playfully before becoming more serious. "So our ball is from sometime between 1920 and 1975. Well, that rules out a bunch of things. Like the madness couldn't have been from the Battle of the Brandywine; that was way earlier."

"Yeah, and it's not the Lenape Indians, either," said Wyatt. "Since most of them left before 1920. And that Hagley explosion was in 1890."

"What about the Underground Railroad and the Civil War? Those were in the eighteen hundreds, so they couldn't be it. Or Zacharia Walker? No, that was in 1911." Danni told Wyatt the story her mother had told her.

"There's been a lot of madness in this area," Wyatt said. "It's kind of sick."

"Yeah, but obviously none of it is *the* madness," said Danni.

Luke and Meg were in the barn talking to the farmer when Wyatt, Danni, and Duke returned from playing in the fields. The barn was cool and windowless, so it took a few minutes for their eyes to adjust to the dimly lit interior. Lining the walls were shelves, posters and a large message board listing vegetables. Wyatt had never heard of some of them. Along with lettuce, broccoli, garlic, and potatoes were the words daikon, scorzonera, and celeriac.

Check It Out!

Farmer Sam noticed Wyatt reading the board. "That's leftover from last November," he explained. "This season's pick-ups haven't started yet." Sam noted that different vegetables are harvested at different times of the year, from late May through November. "The white board has to be updated each week of the season. It's not like going to the grocery store, where you can get any vegetable you want any time of the year. At the store, vegetables that aren't in season locally are shipped from all over the world. How else would you get summer squash in the middle of winter?"

Wyatt had never really thought about that before, about how he could have strawberries in February, when he knew they didn't grow in this area until June, or apples in March, when the orchard had "pick your own" only during the fall.

"It may be convenient to get what you want any time of year," Sam said. "But that food is definitely not as fresh or healthful as local. And the shipping of those foods is not good for the environment. That's why CSAs have the motto, 'Buy Fresh, Buy Local.'"

"What's a CSA?" Wyatt asked.

"Community Supported Agriculture. Members pay us at the beginning of the season, we grow the food, and the members pick up their share each week. They get fresh food, we get financial help before the crops grow, and we all save open space." Sam gestured outside the barn. "I don't have to sell this property to a developer."

Check It Out!

Luke chimed in. "Wyatt is actually working on an environmental project. Tell him about it, Wy." He sounded proud.

Wyatt didn't really think there was any point in telling a farmer about his mystery tennis ball and woodblock, but he didn't want to disappoint his dad. "It's about water quality," he started. "We're trying to figure out what could have caused some kind of *madness* related to the Brandywine Creek, sometime between 1920 and 1975."

Sam's blue eyes sparkled. "*Madness*? I can tell you about madness. One of the things we do on our farm, besides provide fresh, local food to people, is use organic and conservation practices. And that is directly related to water-quality *madness*."

"What do you mean?" After running around the farm property with Danni and Duke, Wyatt didn't remember seeing any water anywhere.

"Let me show you something." Sam led Wyatt and Danni over to the greenhouse where a limited number of vegetables were grown through the winter. Inside, it was warmer and more humid than outdoors, which explained how vegetables could be grown through the cold season. Danni peeled off her sweatshirt.

He picked up an old apple from the counter and cut out a quarter of it. He held both apple pieces up. "If this apple were to represent the earth, which piece represents land and which represents water?"

"The big piece is water," said Wyatt, "because the earth is three-fourths water and only one-fourth land."

"That's right." Sam set aside the large piece of apple and then cut the one-fourth piece in half. "Half of the earth's land is not available for people to use because it's inhospitable: polar regions, deserts, swamps, mountains."

"So only one-eighth of the earth is for people to live?" Danni said.

"Not even that much," Sam continued. He cut one-fourth out of the one-eighth piece of apple. "This larger piece is land that can't be farmed, because it's too rocky or too steep or the soil's too poor." He held up the sliver of leftover apple. "This is what's left for people to use for farmland and buildings and roads and parking lots."

Try It!

"That's not much," said Wyatt.

"No, it's not," said Sam. He sliced off the peel from the small piece of apple. "And just the topmost layer, the topsoil, is what we use for farming and feeding the seven billion people on the earth."

"But what does that have to do with water quality?" Wyatt asked.

Sam walked over to a table set up with three trays of soil, each angled to form a slope. At the lower end of each tray, a "V" was cut out and a cup wedged under it. One tray had plain soil, one tray was filled with plants and sod, and one tray had furrows pressed across the soil. "We also teach students about responsible farming techniques," Sam said. He handed spray bottles to Danni and Wyatt.

"On your mark, get set, go!" Sam, Danni, and Wyatt each sprayed a tray until soil ran out into its corresponding cup. When Sam called "Stop," they checked the results. The tray with plain soil created a cup of muddy water. The water from the sod tray had much cleaner runoff. The cup of water from the tray with furrowed soil was almost as clean as the sod tray's.

Try It!

Sam explained that, back in the nineteen thirties and forties, farmers lost a lot of topsoil from their farming practices. "Topsoil is not only essential for growing food, but it's only a very small percentage of the earth's surface. And since the world's population keeps growing and the amount of farmland is diminishing, we need to conserve that topsoil. We do that now by using techniques like contour and terrace planting. What do you think the cups represent?"

Wyatt hesitated.

"They represent surface water like our streams, ponds, and lakes, which is where runoff goes," Sam said. "Do you think it's good for all of that topsoil to end up in our water systems?"

Danni jumped in. "No, because it makes turbidity," she said, excited to interject the scientific term. "And it's bad for the macroinvertebrates. It can clog up their gills."

Wyatt just looked at her before speaking. "So, what you're saying is that back in the nineteen thirties and forties, farmers were losing topsoil, and the Brandywine Creek was getting bad because that topsoil was getting into it?"

"That's one of the reasons for poor water quality back then, definitely. As a farmer," Sam said, "I'd call that *madness*."

~ ~ ~ ~ ~

The Nystrom's drive from Maysie's Farm to the Earth Day celebration was short, but noisy. As they drove around a woodland, Luke slowed the car. Strange electronic-sounding beeps came from the woods and grew to a deafening cacophony. Thousands of spring peepers, visibly imperceptible, screamed out for mates, their bubbled throats amplifying their calls to an ear-splitting racket.

"Good lord, close the windows," Meg called out above the din.

The car quieted as the windows closed.

"It's only the boys that make that sound," said Danni. "Then the girls can choose who they like, mate, and lay eggs. After the spring peepers lay their eggs, other types of frogs can take their turn looking for mates. Different frogs like to lay their eggs in water of different temperatures. That's so they don't all need to compete for the area at once."

Check It Out!

"How do you know all that?" Wyatt asked.

"From that field trip I went on with Quinn," she said.

~ ~ ~ ~ ~

A banner hung at the entrance to the parking lot of Marsh Creek State Park. HAPPY BIRTHDAY, EARTH DAY! Under the sunny sky, Marsh Creek Lake sparkled. Exhibits and booths flanked the lake. As soon as Luke parked, Danni and Wyatt jumped from the car and ran to greet Dr. Flo at her exhibit for the University of Delaware.

Map It!

"I brought you something," Dr. Flo winked at Wyatt. "Look under the table."

Wyatt pulled the tablecloth aside and found Jennie and Rob huddled under the table. "Surprise," they yelled, laughing and rolling out onto the grass.

On the deck overlooking the lake, a band started warming up. "Ladies and gentlemen, please welcome the Uppatinas Green Shoes Band," the announcer bellowed. The four friends watched as the all-girl band took the stage and started playing a familiar song to a funky rap beat.

"The music's good, but the song is so lame," Rob said.

"No, wait," said Jennie. "Listen to the words."

> *"The itsey bitsey spider climbed up the forest tree*
> *She spun a web to catch something to eat*
> *Some strands were tight and some of them were loose*
> *So she could paralyze her prey and then suck up all its juice."*

Rap It!

"Yuck," Jennie shrieked before hushing them again to hear the rest of the song. The rap beat continued as the lead singer started the final verse.

> *"The itsey bitsey spider now pleasantly was fed*
> *She went to find a male spider to wed*
> *Once she found the male with whom she liked to mate*
> *She got what she wanted and then that male she ate."*

"Ugh," Rob and Wyatt blurted as Danni and Jennie high-fived each other, laughing.

Mixed aromas of funnel cake, popcorn, and hot dogs filled the air. "Speaking of eating, I want to get something to eat," Rob said.

"Yeah, me, too," said Danni.

"I want to check out the displays," Wyatt said, turning away.

Rob turned to Wyatt. "You're such a geek."

Jennie intervened. "I'll go with you, Wyatt," she said.

"Suit yourself," Rob answered. "C'mon Danni."

Jennie followed Wyatt. He passed Dr. Flo's exhibit, where she was talking to a young family and puppeteering a large stuffed grasshopper. "How far can you jump?" he heard her ask. He stopped to watch the children jump about two feet from where they started. "A grasshopper," Dr. Flo explained, "has so many muscles in his legs, he can jump up to twenty times the length of his body."

Try It!

Dr. Flo looked at Wyatt and Jennie. "Check that one out," she called, pointing to an exhibit from Franklin and Marshall College.

On the way, they passed a group of high school students manning a booth titled BRANDYWINE FLOWS: THE FUTURE, THE LEGACY OF OUR WATERSHED. One of the teenage girls called out to them. "Come test yourself on bird and frog calls."

Jennie stepped forward to guess what the bellowing sound was that was coming from a small handheld gadget. "A cow?" she guessed.

The girl laughed. "That's what everyone says," she replied. "But really, it's a bullfrog, the largest frog we have in our watershed." A boy dressed in goggles and flippers stood beside the display table. He manipulated an untied balloon near his throat to make high-pitched squeaks.

Try It!

"He's calling for a girlfriend," the girl said, pointing at the balloon. "Are you available?"

"No," Jennie blurted. She looked at Wyatt, blushing. "I'm so embarrassed that I thought it was a cow," she said. Wyatt hoped that was not really why she was blushing.

They passed several other exhibits. On display for the Brandywine Valley Association was a large map entitled RED STREAMS BLUE. Wyatt paused in front of the map and noticed a notation under the logo: BVA, EST. 1945. A gray-haired man addressed him. "Welcome to the Brandywine Watershed. The blue streams on the map indicate good-quality water. The red streams don't meet water-quality standards, and that's a problem. We at the Brandywine Valley Association are trying to restore these red streams and make them blue again."

Check It Out!

"Wyatt, Jennie," Danni called, interrupting their conversation and gesturing at them frantically. "Come quick! You won't believe this."

Wyatt rolled his eyes. He really wanted to check out the Franklin and Marshall College exhibit, but he could come back later. They scurried over to join Rob and Danni, who shushed them so they could hear about Marsh Creek State Park.

"There used to be a town down there," Danni whispered. "Before it was a lake."

The park ranger, Colleen, was telling the story of Marsh Creek Lake. "As you walk along the trails of Marsh Creek State Park, be on the lookout for ruins of old buildings. Some remain, dilapidated and crumbling with years of neglect. They are just part of the story of Milford Mills, once a bustling little hamlet that became what some call a *drowned town*."

"Drowningtown," a boy called out. The audience laughed nervously.

Jennie shuddered. Wyatt figured she was imagining children and mothers screaming with desperation as the waters rose around them. It was now or never. He reached his arm around and rested it on her shoulder. Jennie offered no resistance.

"Now, I know that some of you assume the worst when we say 'drowned town,' but it really wasn't that horrible," Colleen continued. "No one actually drowned. Marsh Creek, a large tributary to the East Branch of the Brandywine, ran through the center of Milford Mills in this valley." She gestured across the lake that sat like a filled bowl, surrounded by picturesque rolling hills. "Back in the seventeen hundreds, the water power from Marsh Creek ran a grist mill, followed by various other mills through the eighteen hundreds. Eventually, the mills shut down and the property became desolate."

Milford Mills, Wyatt thought. *Downingtown was called Milltown; this was Milford Mills. There were so many mills along the Brandywine.*

"In the nineteen thirties, a Philadelphia gangster purchased the property as a retreat. After he died, the property was sold off. Eventually, it became the little town of Milford Mills, and was made up of a few dozen homes, barns, and buildings. When it was decided that a dam needed to be built to prevent flooding and supply water for the communities downstream, these properties were condemned. The families had to move away."

Check It Out!

Wyatt imagined someone coming to his home with an announcement that they needed his building for some reason, and he and his family had to move, no discussion. He couldn't contain himself. "That's not fair! People lived there."

Colleen hesitated. "Many people would agree with you, that it wasn't fair to the property owners. But it was a matter of *eminent domain*, when the rights of the many override the rights of the few. The State of Pennsylvania deemed it more important to construct a lake for the sake of the people downstream. They have the right to take away private property for a greater public use. Of course, the state had to compensate the property owners."

"But how can you put a price on a community?" a man in the audience asked. "What about the neighborhood? What if you don't want to move?"

Colleen answered slowly. "Unfortunately, that's how it works. But now, we have a seventeen-hundred-acre park with a beautiful five-hundred-acre lake for recreation, downstream flood control, and water supply. The lake holds four million gallons of water and is home to large-mouth bass, walleyes, tiger muskies, and perch, as well as kingfishers, great blue herons, ospreys, and lots of other birds and wildlife. You can boat here, picnic, fish, hike, or hunt. The people of Milford Mills had to make that sacrifice so the rest of us could benefit."

Danni looked puzzled. "How do you make a lake?"

Colleen smiled. "Well, after condemning the town of Milford Mills, moving everyone out and razing the buildings, a dam was built to stop up the creek." She pointed to the flat hill-like structure at the south end of the lake. "Between the creek water and precipitation, it took about seven months to fill up. So, now that you've got a little history of the lake, and, if there are no more questions, feel free to go boating. No rental fees today in honor of Earth Day!"

~ ~ ~ ~ ~

The canoe scratched along the sandy beach as Jennie pulled and Wyatt pushed.

"You're supposed to lift it so it doesn't scratch," Rob said. He lifted the stern of his boat as Danni struggled to lift the bow and carry it to the water.

"It's easier just to drag it," Jennie said to Wyatt.

They launched the two boats and paddled out toward the center of the lake, occasionally bumping into each other.

"We'll race you," Rob called as he pulled his paddle through the water, creating little whirlpools behind him. Overpowering Danni, who paddled laboriously, he J-stroked to straighten the canoe out.

"What are you, a canoe expert?" Jennie called, as she and Wyatt veered off course.

"I have my talents," Rob answered. He turned to Danni. "I actually canoe a lot. My mom taught me how from her camp counselor days."

When they reached the center of the lake, Danni stowed her paddle, turned around and leaned back on the bow, sunning herself. They waited for the remedial canoeists to catch up.

"Jennie Lee, Jennie Lee," Rob called as the second canoe approached. "We're under the sea! Help me, help me, Jennie Lee, Jennie Lee."

Jennie pulled her paddle from the water and leaned over the side of the canoe to peer into the dark water. "That is so creepy," she said. "That there was once a whole town down there."

"Maybe that gangster is still down there," Rob said. "Waiting for company."

"Stop it," Jennie yelled. She angled her paddle just over the surface of the water, threatening to splash Rob.

"Allow me," Wyatt said. "I may not be a great canoeist, but I do know how to use a paddle." He swung, sending a spray of cold water directly at Danni and Rob.

Danni shrieked. She whacked the water in return. Then all four kids splashed furiously in a head-on attack. They swung their paddles and the canoes rocked recklessly. Squeals of surprise, taunts, and good-natured threats wafted over the lake. Wyatt and Jennie both lunged to one side for an assault when, suddenly, their combined weight tipped the canoe, depositing them into the lake. The boat filled with water. Their paddles floated by. Laughing and coughing up water, Wyatt and Jennie, buoyed by their orange life jackets, swam clumsily toward the other boat.

"Don't you dare," yelled Danni, afraid of retribution now that her cousin had nothing to lose.

Rob yelled. "Get your paddles and stay with the canoe. We'll come to you."

Wyatt and Jennie turned awkwardly, resembling a pair of orange manatees. They retrieved the paddles and stowed them under the thwarts of the submerged canoe. Following instructions, they turtled the boat and pushed it perpendicular to Rob's. "Danni, lean the other way," Rob said, taking full control of the rescue. He leaned over the gunwale of his boat and lifted one end of the overturned canoe slightly to break any suction between boat and water. Then he heaved the bow onto his gunwale, draining the water out. He and Danni pulled the boat across theirs, forming a giant canoe "X," turned the canoe right side up, and pushed it back into the water.

Jennie and Wyatt grabbed onto opposite sides of the restored canoe. Wyatt leaned heavily on his side, countering Jennie's weight as she hoisted herself up and flopped into the canoe. She leaned away from Wyatt so he could pull himself up and in.

"Oh my God," Jennie said. "I'm freezing!" She hugged herself and shivered vigorously.

"That was hilarious!" said Danni, oblivious to the potential danger of hypothermia.

Check It Out!

Keeping his center of gravity low, Wyatt stepped over the thwarts toward Jennie. When he reached her, he circled his arms around her, thinking he would warm her up.

Rob intervened. "That's not really going to help, Wyatt, since you're wet, too. Both of you need to paddle. That's the only way to warm up."

It was Wyatt's turn to blush. He turned away sheepishly and crawled back to the stern of the canoe. Both he and Jennie raised their paddles and headed toward shore just as the boat patrol sped toward them.

~ ~ ~ ~ ~

The sun streamed in through the car windows, warming Wyatt and Jennie. They sat quietly in the back seat, looking out the windows. Both were wrapped in blankets provided by the first aid station. Outside, the Earth Day celebration continued, though no longer for the Nystroms. Wyatt's parents decided to get Wyatt and Jennie home as soon as possible. Dr. Flo would bring Rob and Danni home at the end of the day.

"Sorry," Jennie finally said after a few minutes.

Wyatt looked at her and they both laughed. "Don't be," he smiled. "You're like an official Polar Plunger now."

Field Trip!

May

Chapter 24

BREAKTHROUGH

Under Uncle Clayt's portrait, a stack of papers lay on Dr. Flo's coffee table. Wyatt looked the portrait over, checking for any changes since the last time he was there. *Same nod, same wink, same smile*, he thought. *OK, no changes this time.*

Wyatt, Danni, Rob, and Jennie had planned to spend the rainy Saturday afternoon at Dr. Flo's, mapping out the data she told them had accumulated throughout the spring. "I need my coffee table back," Dr. Flo had teased.

The large watershed map loomed in front of them. On the table, colored stickers and markers were organized in containers.

"Let me know if you need any help," Dr. Flo said as she breezed through the room. "I'll be in my study making arrangements for my field trip."

Wyatt studied the map. "OK," he began. "Let's just do what we did before. We need to make a sticker for each data sheet and put on it the Biotic Index. Remember, blue stickers are for sites with a Biotic Index greater than ten; that means the water is good. Red stickers are for sites with a Biotic Index less than ten; that's not so good."

Just as they had done several months earlier, the friends labeled and applied stickers. Wyatt read information from the data sheets, Jennie recorded it on the stickers, and Rob and Danni located and placed the stickers on the correlating map sites. From the confluence up, blue and red data stickers dotted the East and West Branches of the Brandywine and their tributaries.

Red Stickers	Blue Stickers
West Branch Brandywine at Embreeville; BI=9	Broad Run; BI=11

West Branch Brandywine at Mortonville; BI=7	Doe Run near Route 82; BI=12
West Branch Brandywine at Modena; BI=5	Buck Run near Route 82; BI=14
Sucker Run; BI=5	Rock Run; BI=12
West Branch Brandywine at Route 30; BI=4	East Branch Brandywine, Glenmoore; BI=12
Shamona Creek; BI=7	East Branch Brandywine, Lyndell; BI=14
East Branch Brandywine just below Downingtown; BI=8	
Valley Creek by Exton Mall; BI=6	
East Branch Brandywine at Kerr Park, Downingtown; BI=7	
Beaver Creek near Downingtown; BI=6	
East Branch Brandywine just below confluence with Shamona Creek; BI=7	
Taylor Run near West Chester: BI=7	

Sticker It Red or Blue!

"Dr. Flo, come see what we've got," called Wyatt, as the four friends surveyed the map.

Dr. Flo appeared in the doorway, a large manila envelope in her hand. "Wow! Your network of schools has really covered a lot of the watershed. What have you found?"

They reviewed the data before them, making their way up and down the East and West Branches of the Brandywine and their tributaries.

"Mostly red," said Wyatt. "That's what we found."

"And most of the blue stickers are on tributaries. There are only a few on the Brandywine," added Danni.

Dr. Flo smiled. "Bravo!" she said. "Good work. And it looks like you're in luck. Next week, my graduate-level class will be running water-quality tests on the upper portions of the watershed, where the highest headwaters of the Brandywine are. They are going to compare their results with what they found on their last field trip in the lowest parts of the Brandywine, down in Wilmington." She paused for a moment, a glint in her eyes. "I wonder if I can get you kids out of school for a day to come along with my class. Would you be interested?"

"Are you kidding?" Rob said. "Miss a day of school? Oh, yeah." All four kids nodded enthusiastically.

Dr. Flo looked sternly at him. "It won't be fun and games. It'll be work, and you'll be expected to pitch in."

"Yeah, but at least it's work that makes sense," Wyatt said. "Count us in."

Dr. Flo turned to Wyatt. "Great. By the way, I saw that you never made it over to the F and M display at the Earth Day event, due to your little *accident*." Danni giggled. "So I picked up some materials from them. I thought you might be interested." She handed him the envelope.

~ ~ ~ ~ ~

The dreary evening passed slowly. Meg and Luke Nystrom had gone out, leaving Wyatt on his own. He sat down at the kitchen table with a bowl of ice cream, ready to get to work. He opened the envelope from Dr. Flo. Maps and papers, packed with data and information, spilled out across the table. Wyatt picked up a map. *Dams of the Brandywine.* It was illustrated with scores of little red triangles along the imperfectly shaped *Y* of the Brandywine Creek and its tributaries. *Almost one hundred thirty dams*, he noted. Another paper listed the types of mills. Gristmills, clover mills, cotton mills, iron forges, saw mills, paper mills, and rolling mills all, at some point, made use of the Brandywine for power. Picturing the Hagley Museum's gunpowder mill, Wyatt tried to imagine a hundred other mills along the Brandywine and its tributaries, each with a dam backing up and channeling water into a millrace to power a water wheel and provide energy. That was a lot of mills on the creek.

An article caught his attention: "Legacy Sediments a Breakthrough." Wyatt skimmed through the parts about southeastern Pennsylvania's mill history. The article read:

> *Each dam slowed the stream's flow nearly to a halt, creating a slack-water pond. Through the centuries, sediments from the cleared upstream woodlands*

accumulated behind these dams. Though the mills have long stopped functioning, most of the accompanying dams have decayed and breached. The water, once held back, flows freely again. Unfortunately, those free-flowing streams slice through those "legacy sediments," some of which are ten feet high, eroding and carrying them downstream, where many end up in the Delaware Bay. This recent discovery attributes much of our water-quality issues regarding erosion and sedimentation to our forefathers, rather than to present-day development and farmland practices. According to Professors Merritts and Walter, these legacy sediments will haunt our waterways unless we restore them to their pre-settlement structure of meandering wetlands.

A photo accompanied the article. Two people, identified as Franklin and Marshall College professors Dorothy Merritts and Robert Walter, stood in a creek bed wearing waders and measuring the six-foot-high stream bank beside them.

Wyatt closed his eyes, trying to picture what the Brandywine might have looked like long ago, before the landing of the Swedes at the mouth of the Brandywine back in 1638. A series of meandering wetlands and roaring waters. "Where do you think Pocopson school got its name?" he recalled his mother asking. *The land was cleared by colonists so they could build homes and farms,* he thought. *But that destroyed the hunting grounds for the Lenape Indians. And it sent lots of soil into the streams. Then the colonists built mills with dams that backed up the water behind them. And that screwed up the fish that needed to migrate, so the Lenape lost their fishing grounds. Then the mills and more modern factories sent pollutants, chemicals and dyes into the water. And then the farms of the 1940s sent tons of topsoil and pesticides and herbicides to the streams. And factories and treatment plants put their wastes into the water before there were laws against doing that. And then there's nonpoint sources of pollution from building neighborhoods and strip malls and roads. And now legacy sediments from those dams from the 1800s? It's ALL madness!*

An Indian woman approached Wyatt, her long gray braid draped over one shoulder. She looked at the table covered with papers, and picked up the ice cream bowl that was next to Wyatt. Tilting the bowl to her lips, she slurped the melted ice cream. "*Welhik*," she said, wiping her mouth with the back of her hand. "*Wanishi*."

Try It!

Wyatt just stared. The woman sat at the table across from him and placed the bowl down before speaking. "Do not inflict madness upon this earth as was inflicted upon my people," she began in her slow monotone. "You have done your work well," she smiled. "*Now* is the time of the Fourth Crow."

~ ~ ~ ~ ~

Wyatt pushed away the hand nudging his shoulder. "Wyatt, wake up," his mother said gently. "You fell asleep at the kitchen table."

Wyatt looked up, his eyes blurry from sleeping on his crossed arms. "Go to bed, kiddo, it's late."

He rose from the table. "Can I clean this up tomorrow?" he asked, seeing that his papers were sprawled across the table.

"Sure," his mom replied. "I'll just put this in the sink to soak." She picked up the ice cream bowl that sat on the table directly across from where Wyatt had fallen asleep.

Chapter 25

SWEET WATERS

Dr. Flo turned the van east. The early morning sun glared, so she flipped down the sun visor. As she turned north on Reeceville Road, the sun streamed in on the passenger side of the van. Wyatt, Danni, Rob, and Jennie sunk down in their seats, trying to avoid the strong sunlight. The graduate students in the van didn't seem to mind, and chattered away, sipping cups of steaming coffee.

"We're taking a short detour to pick up the rest of my students," she said, too cheerily for the early hour. "No sense in them driving down to Wilmington if we're headed up this way anyway."

A roadside sign welcomed them to WEST BRANDYWINE TOWNSHIP, BETWEEN THE BRANDYWINES. "Now, at which school did they say they'd be?" Dr. Flo said. Slowing down, she scanned the outside of the schools that were lined up along the road: Reeceville Elementary, North Brandywine Middle School and, finally, Friendship Elementary, where three teachers stood at the main entrance.

Map It!

"Everybody out for a minute," Dr. Flo called to the van of students. "I want you to see this."

The adults clamored out of the van, followed by the slower-moving Wyatt, Danni, Rob, and Jennie. The three teachers joined them.

"They look like teachers, not students," Rob whispered to Wyatt.

"Well, my mom's a teacher, but she's getting her master's degree, so she's a student, too," Wyatt explained.

Dr. Flo cleared her throat to get everyone's attention. "Before we move on, I wanted you to take a look around." In the distance, a groundhog lumbered across a field and disappeared into a hedgerow. Chickadees chittered and flitted from tree to tree. Bluebirds swooped across the school property. A brilliant-red cardinal chased a duller female. Sitting on a wire, a mourning dove sang, sounding eerie and owl-like.

Check It Out!

"We are lucky to have such a glorious day," Dr. Flo said. "Now, look to the east." The group turned toward the sun rising over a cluster of homes. Wyatt followed their gaze. "Everything that you see in this direction is part of the East Branch Brandywine watershed. If you look to the west, you will see the West Branch watershed. We are standing at the boundary between the two." She pointed to the sign on the road: BETWEEN THE BRANDYWINES.

Check It Out!

Wyatt looked from east to west and back again, finding it hard to see that they stood at a highpoint. But he didn't doubt Dr. Flo. "We'll be testing on both branches today," she continued, "traveling up the West Branch first and then down the East Branch. We have a long day ahead of us, so let's get going."

Farms, old ranch homes, orchards, and new cookie-cutter neighborhoods bordered the road to Hibernia County Park. All around them, spring was in bloom. Clusters of white flowers hung from locust trees like bunches of grapes. Azaleas flashed their pink, red, and white blossoms.

Map It!

Dr. Flo entered the park from the south end, crossing the West Branch of the Brandywine. "We'll be testing here in a little bit," she called back from the driver's seat. "But first, we'll head up to the lake."

A few cars were parked in the lot at Chambers Lake, most likely owned by the fishermen in boats scattered around the reservoir. "The lake is a wonderful place for recreation and wildlife," said Dr. Flo.

"Is that a turtle?" Rob asked, looking through the van window. By the side of the lake, a wheel-sized snapping turtle lay still, her pointed beak and long, serrated tail distinguishing it from other kinds of turtles.

"Love is in the air," Dr. Flo answered. "Or at least it was. It looks like we have a female on her way to laying her eggs."

Check It Out!

Dr. Flo explained that Chambers Lake was created by damming up Birch Run. Hibernia County Park had been a thriving iron plantation in the 1800s and had dams on it to provide water power for the mill. "The new dam serves a different purpose. It creates additional water supply for Coatesville and helps control floods."

"Was there a drowned town?" Wyatt asked. "Like at Marsh Creek Lake?"

"No, not here," said Dr. Flo. "Since it was already park land, there was no need to condemn any residences or businesses. Chambers Lake was a lot less controversial than Marsh Creek Lake." She then explained that they would head to the first test site of the day. "We'll be going to the West Branch of the Brandywine, just below this lake and the confluence with Birch Run."

Map It!

She turned the van around and drove down the road to a spot by the West Branch. The group unloaded equipment and set out for the test site. They crossed an old bridge and walked along a carriage lane that bordered the creek. Surrounding them was forest, cool and damp. Just a few beams of sunlight filtered through the tree canopy. Small, secretive wildflowers dotted an understory of mountain laurel, rhododendron, and ferns. In the creek, water tumbled between boulders. *Roaring waters*, Wyatt thought.

Dr. Flo led the group downstream to a milder riffles area. They divided up the chemical, biological, and physical testing equipment, and got to work.

~ ~ ~ ~ ~

Before leaving the park, Dr. Flo drove the van down a gravel road, just past the striking yellow-and-red Hibernia mansion. Wyatt climbed from the van to get a closer look at the ruins of the iron mill and a diagram outlining the iron mill process. A stone structure with three slots, once hosting water wheels, stood empty, and the milldam had crumbled long ago. The entire place reminded Wyatt of the Hagley

Museum, where they used water power to make gunpowder. If he hadn't seen that preserved mill, he would have had a much harder time picturing this one in its working form. He imagined water channeled through a millrace and then rushing through the slots, forcing the water wheels around. The diagram showed one water wheel driving bellows that pumped air into a furnace. From above the furnace, men dumped in iron ore, lime, and charcoal. The charcoal burned, heating the mixture until the lime bonded with the nonmetallic parts of the iron ore to form a removable slag. That left pure iron to trickle out of the bottom of the furnace, forming bars of commercial iron.

Check It Out!

Dr. Flo interrupted Wyatt's thoughts. "The birth of steel ended the days of the iron forges. Steel was stronger and more workable. So this iron plantation became obsolete. The property became a country estate and, eventually, this lovely county park."

Wyatt thought of something else. *The dam. That's where the sediment would have built up.* He pictured it breaking and the Brandywine gushing free, carrying loads of legacy sediment downstream.

~ ~ ~ ~ ~

The van slowed as a horse and buggy turned onto the road ahead. The clip-clop of horse hooves produced a comforting rhythm, though Wyatt heard Dr. Flo's sigh of relief when the buggy turned off the road into a gravel driveway. The van struggled up the incline to the Baron Hills, hit the crest, then charged down the other side, where a valley opened up in front of them. Large stretches of farmland filled the landscape, some ending abruptly at the boundary of a new housing development. Dr. Flo parked on the side of the road. A boy in a straw hat and dark trousers passed them on his scooter.

Check It Out!

In front of them, a small creek crawled through the pastureland. Beyond the creek, a white barn loomed over a muddy field. Scores of cows, their hooves sunk into the mud, stared glassy-eyed toward the van. They swished their tails slowly.

"Our next test site, everyone," Dr. Flo said, gesturing toward the creek.

"*This* is the Brandywine?" a woman asked.

Dr. Flo turned to look at her students. "The beginnings of the West Branch," she said.

Map It!

Rob whispered to Wyatt. "This is the same West Branch as the one we tested in Hibernia?"

"Yuck," said Danni. "It's probably full of cow poop."

"I think they're called cow pies," said Jennie.

"*It starts with an 'S,' ends with a 'T,'*" sang Wyatt.

Rap It!

Dr. Flo pointed to the farmhouse behind the barn. A clothesline full of dark-purple dresses and black pants fluttered in the breeze. "You can tell it's an Amish farm," she said, "not just by the clothes on the line. Take a look at the power lines."

Wyatt looked along the road. None of the power lines connected to the farmhouse. Unlike the ranch homes and mobile-home parks nearby, this property had no cars, no tractors or trucks, no satellite dishes. Off in the distance, a set of horses pulled a plough through the field, churning the soil. A lone windmill turned slowly, close to the farmhouse.

"The Amish live self-sufficiently as a community, independent of the wider world," Dr. Flo said. "Many of them are farmers and are concerned about the impact their farms have on the Brandywine. You can see that the creek here is completely vulnerable; there are no riparian buffers. But some of the farmers have put into place some *best management practices,* like contour farming." She pointed out the alternating rows of corn and wheat in the fields. "And containment systems to keep the cow manure from washing into the creek."

Check It Out!

"I don't even want to know how that works," Jennie whispered to Wyatt.

"That's full of cow manure?" Wyatt said, pointing toward a low, circular building. "Holy scat!"

They unloaded the testing equipment and divided it up. The Welsh Mountains to the north and the farms surrounding them were picturesque. But Wyatt wondered what kind of water-quality results they would get here, compared to the results done in the cool, forested creek just a few miles downstream.

~ ~ ~ ~ ~

Just down the road, they crossed a small hill. They drove a little farther when Dr. Flo parked the van again. "We are now on the East Branch," she said. "That last hill was the divide between the East and West Branch Brandywine watersheds." Though there were no cows eyeing them and no barn surrounded by muck, this site was also bordered by open fields with little protection. No trees, no shrubs, no shade cooled the waters of the creek from the intensifying sun. Again, it was hard for Wyatt to imagine this humble beginning of the East Branch of the Brandywine as part of the great creek that provided power to grow and develop the many industries, communities, and businesses downstream. He looked up and saw a pair of crows flying overhead. The words came back to him from the night of the winter solstice celebration. *The second crow tried to clean the world, but he became sick and died. The results of this testing,* Wyatt thought, *probably won't be so good.*

Map It!

~ ~ ~ ~ ~

Struble Lake sparkled in the sun. Wyatt sat on the grass with Jennie, Rob, and Danni. "I'm so hungry," he said. "I feel like it's been days since I had breakfast." He pulled a peanut butter sandwich from his lunch bag.

Rob stared at the sandwich. "Are you going to eat that whole thing?"

"Don't tell me; you forgot your lunch. Here," Wyatt said, handing half of his sandwich over. Following Wyatt's lead, Danni offered Rob some of her chips, and Jennie broke her granola bar in half.

Fluffy milkweed seeds floated by in the light breeze. By the boat ramp, a thicket of cattails stood, some tall, some bent under the weight of red-winged blackbirds resting on the stalks. Phragmites swayed gently. From a box mounted nearby, a bluebird emerged. It flew overhead and sunlight reflected its cobalt-blue wings and orange throat. A fishing boat approached the ramp, its electric motor quiet. The fisherman cut the motor and let the boat drift toward the ramp. He jumped off and pulled the boat to shore, then walked toward his truck.

"Florence?" the fisherman said, as he passed the picnicking group. "It's been a long time."

Dr. Flo turned around. Her eyes lit up and she smiled at the tall, lean man with the scruffy mustache. "Abe, what a surprise seeing you, although I should have known you'd be out here fishing on such a glorious day. Some things never change."

"Well, if they build a lake in my backyard, you can bet I'm going to take advantage of it." He stooped, since she was at least a foot shorter than he was, to give her a hug.

Dr. Flo chatted with Abe for a bit before he turned toward her students. "And these would be your students?" he said.

She explained the nature of the day's trip and the water-quality testing they were conducting. "We have one more site to test downstream on the East Branch at Springton Manor Park. Right now, we're taking a well-deserved lunch break."

"Why did they build this lake?" Wyatt asked. "When they have Chambers Lake in Hibernia?"

"I guess you need that map in front of you, Wyatt," Dr. Flo said. "Chambers Lake provides water supply and flood control on the Brandywine's West Branch. Struble Lake does the same, but for the East Branch, along with Marsh Creek Lake. And of course, it provides recreation, too."

Map It!

Abe eyed Wyatt. "You seem pretty young for a college kid."

Before Wyatt could answer, Dr. Flo interceded. "They're just playing a little hooky to get some real world schooling."

Abe turned back to Dr. Flo. "I'm really glad I ran into you. You know, my pop died last year at the ripe old age of ninety-eight. We've been cleaning out the house and came across some things that I think belonged to Clayt. I haven't been able to get a hold of his family, so I figured you'd be the next best person to contact. You'll have to come up and visit. I think we've got some stuff you'll want to see."

"Sounds intriguing," Dr. Flo said. "I'll give you a call and come up as soon as exams are over and grades are in."

Several of Dr. Flo's students groaned. "No hurry on that," one of them said.

~ ~ ~ ~ ~

WELCOME TO WEST NANTMEAL read the sign on the road leading away from Struble Lake. LAND OF THE SWEET WATER.

Map It!

"The sweet water of the Brandywine," Dr. Flo commented.

It sure didn't look sweet at the last two places we tested, thought Wyatt.

As though she had heard his thought, Dr. Flo said, "It wasn't too sweet in Honey Brook, though, was it? But think about it. *Honey Brook*. It must have been sweet water there once, too."

"Honey brook, sweet water, oh, I get it," said Rob.

Dr. Flo continued. "And just you wait. It'll be sweet again soon. That's the thing about the creek. There are so many influences, good and bad. The creek will be impaired in one place, like Honey Brook, and exceptional somewhere downstream, all because of what? Wyatt, you should know the answer to that."

Wyatt thought back to the large map posted on Dr. Flo's wall and the boundaries he and his friends had outlined. He recalled lying down and extending his arms with finger tributaries. "Because of the watershed," he said. "Because of the land that drains into the creek and because of what the tributaries add to the creek."

"Right," Dr. Flo said. She winked at him in the rearview mirror. "The area around the creek really influences its quality."

The rural road through the forested landscape made it feel as though they were in the middle of nowhere. Wyatt noticed a stone wall paralleling the road for miles. It seemed comforting, as though it were guiding them somewhere. Ahead, the trees thinned out. The sky broke open and suddenly, looming above them, a castle appeared. Its salmon-colored roof shone against the blue sky. Pointy spires, turrets, and deep arches of the gothic-revival mansion were magnificent. *On a stormy day, this place would definitely give me the creeps,* Wyatt thought.

"I remember when I first saw this place," said Dr. Flo. "I used to come up here with Uncle Clayt sometimes to explore the hidden nooks and crannies of the Brandywine Valley. I thought this place was magical; I wanted to live here."

"What is it?" Danni asked from a few seats back in the van.

"It was originally called *Langoma* and was a home of the Potts family. You've heard of Pottstown and Pottsville, haven't you? The Potts family started an iron-making business and ran the Isabella Furnace down the road. They built this mansion that is now a retirement home run by the Daughters of St. Mary's of Providence."

Map It!

"What's the Isabella Furnace?" asked Jennie. "And why would they build it here in the middle of nowhere?"

"Good questions. Hold on a minute," said Dr. Flo, "and I'll explain."

She continued driving through the wooded landscape. A gated driveway, bounded by stone pillars, appeared on the right. She pulled over. ISABELLA FURNACE. At the top of the driveway, a towering stone building stood. Once a part of the iron furnace, it was now an impressive residence.

"The iron furnace was located here for several reasons. Can you guess why?" said Dr. Flo.

The stream nearby was a clue. Wyatt nudged Danni and pointed to Perkins Run. "Water power?" Danni guessed.

"Absolutely," Dr. Flo said. "So much history of the Brandywine Valley is due to the power that water provided for the mills and furnaces. What else? Think *furnace*. What does a *furnace* do?"

"Burns wood," Jennie said hesitantly, and then perked up. "The forest and trees! There's lots of firewood here."

"That's right," Dr. Flo said. "They felled trees to make charcoal to burn in the furnace. So they've got the fuel for burning and the water for power. What else did they need?"

"Iron?" Rob said.

"Bingo," said Dr. Flo. "There was iron ore in the Welsh Mountains, where the headwaters of the Brandywine are. They mined it, as well as all these stones you see lining the road. When work was slow, Colonel Potts had the workers build rock walls to keep them working."

Dr. Flo gave an imaginative description of what the community was probably like in the 1800s. Wyatt closed his eyes and pictured the images she conjured: the wealthy iron master and his family living in his fine home on this two-thousand-acre estate; cooks and housekeepers bustling around the house, attending to daily chores; children and wives of the mill workers sowing seeds and weeding in the vegetable gardens; sweaty, muscular men felling trees and slow-burning the wood in pyres to make charcoal; the orange glow of the iron furnace, its acrid smells wafting through the air, and charcoal ash darting across the skies; teams of mules pulling loads of stones on the roads from the Welsh Mountains; the water of Perkins Run, rushing on its way to the East Branch of the Brandywine, turning the wheels of the mill that kept this miniature empire going.

Check It Out!

"Hey, daydreamer," Jennie teased, poking Wyatt in the ribs. "Time to wake up."

Wyatt hadn't realized that Dr. Flo had been driving again, and had stopped to point out another feature along the creek. Boulders and pine trees separated their parking spot from a large field with a huge hump of a hill nearby.

"Time for a little football," said Dr. Flo. Wyatt and Rob perked up, so she quickly added, "Just kidding." She proceeded to point out the breast of the Barneston Dam, the hump. "It's a dry dam," she said.

"It only holds back water in times of flooding. Otherwise, the creek can flow normally through a hole in the dam. If there's too much rain, the dam will hold back the water so it won't flood downstream. The football field is an ample spillway, so if the water gets too high, it won't flow over the top of the dam, which could destroy it."

"Can we play football? Just for a break?" Wyatt asked.

"No time for that, kiddo. We have one more stop to make, and then we'll call it a day. I told you this was not going to be fun and games, but real work."

The van wound its way down Route 282, along the Brandywine's East Branch. Tall sycamores shaded the creek. Water tumbled around rocks, looking much more like the Brandywine's West Branch in Hibernia County Park than like the East Branch up in Honey Brook. In the historic hamlet of Glenmoore, they turned onto Indiantown Road.

"Somewhere around here is an Indian burial ground," Dr. Flo commented. "We'll have to come up and explore some day when we have more time."

They drove to the main entrance of Springton Manor Farm and down the long driveway. Before getting out of the van, Dr. Flo told the group she had a special treat. They all ambled down a path past the grand Springton Manor house and a Victorian garden bursting with purple irises and white peonies, and arrived at a large, screened-in building. Inside, a garden brimming with coneflowers, milkweed, bee balm, and black-eyed Susans surrounded a small pond. Fluttering from flower to flower, butterflies colored the space: monarchs, swallowtails, zebras, fritillaries, hairstreaks, and skippers. The adults dispersed to explore the space. The kids stayed with Dr. Flo.

Map It!

"It's simply magical in here," Dr. Flo said, examining a jewel-like chrysalis suspended from a twig. "Can you imagine what's going on in there?"

Danni and Jennie looked closely at the chrysalis. "Not really," said Jennie.

"Inside of that chrysalis, a caterpillar is breaking down completely before reshaping itself into a monarch butterfly." The girls looked skeptical. "Did you ever see the movie *Willy Wonka and the Chocolate Factory*? Not the remake with Johnny Depp; that was unfortunate. But the original, with Gene Wilder."

"I love that movie," Danni said.

"Do you remember the scene when Mike Teevee gets broken into millions of little pieces and gets sent through the transporter machine?"

"Yeah, he's so obnoxious, and he jumps into the transporter and gets shrunk," Jennie said.

"Well, that's the same idea as what's going on in the chrysalis of the caterpillar. Its body is broken down into tiny pieces and rearranged. Instead of coming out the same, but smaller, like Mike Teevee, the caterpillar is rebuilt into a butterfly. It's absolutely astounding, another of nature's miracles."

Check It Out!

The group walked down the hickory-lined lane and toward the creek. Sounds emanated from the working farm's barn: cows lowing, goats and horses neighing, ducks quacking. They passed a magnificent oak tree with limbs that trailed along the ground, beckoning to be climbed.

"That oak has been here a long time," Dr. Flo said. "It was around when William Penn was granted the areas now known as Pennsylvania and Delaware. That tree is at least three hundred years old. I'm sure it's seen a lot of history through the centuries." She continued down the trail at a quick pace.

"Why is there so much spit here?" Wyatt asked, observing numerous wads of white goo on the goldenrod and milkweed along the pathway. He thought it might have something to do with the barn animals.

Dr. Flo stopped so suddenly that Wyatt almost bumped into her. She swiped a bubbly mass from a plant stem and gently pushed the bubbles aside, revealing a tiny green creature.

"Cool," Jennie cried. "What *is* that?"

"This little spittlebug creates its own protective covering of bubbles and, like a butterfly, will metamorphose. It will become a full-grown froghopper, an insect."

Several of the adults reached into the grasses and shrubs.

"Please don't," Dr. Flo said. "Let's leave them to grow up as they were meant to do." She lowered her voice so only the kids beside her could hear. "Besides, that's not spit. It comes out of the other end."

Check It Out!

As they approached the woodland, Dr. Flo stopped to point out a tall oak. A cluster of leafy vines surrounded its base and hairy vines climbed its trunk. "Leaves of three, let it be," she warned. "It's poison ivy clinging to this oak tree. Any part of the plant can cause a rash, including the leaves and the roots. Don't touch!" Several birds flitted nearby. "Unless, of course, you're immune like they are. Birds can even eat poison-ivy berries."

Check It Out!

She walked just a few steps forward and stopped again to pluck the translucent stem of a showy orange flower. Using her fingernail, she sliced the stem open and rubbed the juicy interior on her wrist. "The antidote to poison ivy," she remarked. "Jewelweed. I sure wish I had known about jewelweed when I was a teenager. I had to use *nature's bathroom* during a canoe trip and spent the rest of the summer paying for it. I didn't know enough about poison ivy not to squat in it."

Check It Out!

Tulip trees, oaks, beeches, and sycamores towered over Indian Run, forming a canopy that kept the creek in deep shade. The understory of the forest held plumes of ferns, shiny-leafed mountain laurel, and mayapples with white flowers peeking out from under leafy umbrellas. Curious, Wyatt stopped to roll a rotting log. A salamander darted out. Ants panicked, scrambling in all directions, some carrying eggs as they searched for new cover. Centipedes and beetles scattered while a slug simply retracted its antennae. Wyatt picked up a thick millipede to look at its "thousands" of legs. *Licorice*, he thought, getting a whiff of the invertebrate. *It smells like licorice.* He placed the millipede on the ground and rolled the log back into place.

Try It!

The rest of the group, acting like pros, divided the equipment and got to work for the final physical, chemical and biological tests of the day. Wyatt joined his three friends, who had decided to do the biological sampling. They headed over to a riffles area and lowered the seine net into the water, weighting its bottom with rocks, so no critters could slip under. While Rob and Jennie held the net in place, Wyatt and Danni lifted and scrubbed rocks just upstream of the net, dislodging macroinvertebrates from their homes.

"Hurry up," Jennie complained. "It's so cold." The water was cold and the shade kept the air cool as well.

"OK, let's bring the net over to shore," Wyatt directed.

Rob and Jennie removed the weights and lifted the net from the bottom up, making sure not to lose any critters over the top. As they waded back to the bank, the net came alive with dozens of little creatures wriggling in protest of the eviction from their homes. They reminded Wyatt of the people from Milford Mills, the *drowned town*, and he felt bad about disturbing them. *Eminent domain,* he thought. *When the rights of the many override the rights of the few. But we'll put you back, I promise.*

Something large caught his attention. "A hellgrammite," Wyatt yelled. "And a crayfish."

Jennie screamed, but held on to the net.

They placed the net on the ground. Using paintbrushes and cups, they plucked and sorted the critters. Wyatt clasped the crayfish at the abdomen and placed it in a bucket while Danni used a spoon to push the hellgrammite into a cup. Stoneflies and mayflies scurried frantically, searching for cover. Caddisflies and planaria stretched in confusion. Riffles beetles crawled along the net, while water pennies clung motionless. Jennie pushed something that looked like a waterlogged caterpillar into a cup of water.

"What is this?" she said, making a face.

Dr. Flo made her way over to the biology group to see what all the commotion was about. "A cranefly larva," she said. Looking in the different yogurt cups, her eyes widened. "A hellgrammite, stoneflies, mayflies; this is some sweet water!"

"Is this even a bug?" Jennie asked, still looking at the creature in her cup.

"A macroinvertebrate, oh yes," Dr. Flo said. She explained that, just as the caterpillar undergoes metamorphosis to become a butterfly, many stream invertebrates metamorphose into adults. "This cranefly larva," she explained, "will change into what we used to call a *giant mosquito*. Unless, of course, it gets eaten first. To a frog or fish, this larva sure would make a nice juicy meal."

Check It Out!

Jennie looked at Wyatt. "I think I'm going to throw up," she said.

Maybe you'd rather eat the licorice bug, thought Wyatt. But he didn't say anything.

~ ~ ~ ~

The sun had begun to descend in the western sky when the van crossed the border into Delaware. Wyatt figured that he would have gotten home from regular school hours earlier. Although he was tired, he loved spending the long day outdoors. He wasn't sure how Jennie felt about it, though, after that cranefly incident. She had actually seemed nauseous when they got back on the van to head home. He wanted to ask how she was feeling now, but didn't want to wake her. She seemed to sleep so peacefully, her head resting on his shoulder.

Field Trips!

Summer

June

Chapter 26

THUMBS UP

"What the heck," Danni said. She and Wyatt walked toward Dr. Flo's front porch. Although they had both gotten used to the bug sculptures, chipped paint, bat houses, rain barrels and overgrown gardens, there was always a new surprise at the "kooky hoose." Sheets and towels, jeans and shirts and, to their embarrassment, some very large underpants, hung from the clothesline in the yard. While the laundry didn't seem that unusual, the balloons decorating the clothesline and the sign hanging on the porch, CELEBRATE INTERNATIONAL CLOTHESLINE WEEK, took them both by surprise.

Try It!

They knocked on the door and let themselves in. Dr. Flo had told them to treat her home like their own and, after seeing her large panties hanging outside, they figured she meant it.

"Hello," Wyatt called out.

"I'm in here," Dr. Flo responded from her study. "I'll be out in a minute."

As he entered the living room, Wyatt glanced at the portrait of Uncle Clayt. There he was with his slight wink, slight nod, and slight grin. Seeing that Uncle Clayt hadn't changed since the last time he looked, Wyatt headed over to the big map on the wall, which is what he really came to see. He scanned it for the locations where they had tested the creek the week before.

"Hello, hello," Dr. Flo said, entering the room. "I just got off the phone with my friend Abe, and I'm going up to Honey Brook tomorrow to get Uncle Clayt's things. You remember meeting Abe last week at Struble Lake, don't you?" Wyatt and Danni both nodded.

"Would you two like to come with me? I think some of the things may be of interest to us all."

"Sure," Wyatt said. He didn't think he had anything else to do on Sunday, and any day with Dr. Flo usually turned out to be pretty interesting.

"OK," said Danni.

"All right, then," Dr. Flo said. "We'll make a day of it. See if you can be here by ten tomorrow morning." She reached for the data sheets they had completed on their last trip and handed them to Wyatt "Now, I'll leave you two to get this data on the map."

Knowing the routine, Danni took blue and red stickers and labeled them.

Red Stickers	Blue Stickers
Brandywine West Branch at Suplee Road; BI=5	Brandywine West Branch at Hibernia; BI=12
Brandywine East Branch at Suplee Road; BI=6	Indian Run in Springton Manor Park; BI=14

Sticker it Red or Blue!

To place the stickers on the upper part of the map, Wyatt had to drag a chair over and stand on it. Danni handed him one sticker at a time and he placed each on the map until all of the stickers were up. Stepping back to look over the entire map, he missed the back edge of the chair and fell. The map seemed to spin into the ceiling and he glimpsed the portrait of Uncle Clayt who winked, nodded, smiled and gave him a thumbs up. He hit the floor, head first.

Dr. Flo rushed into the room. She helped Wyatt sit up. "Are you all right?"

"Yeah, I'm OK," he answered, though he was a little dizzy from conking his head.

"How many fingers do you see?" Danni tested, making a peace sign with her hand. "*Welankuntewakan,*" she added, giving him a hint.

Ignoring her, Wyatt turned around to check the portrait of Uncle Clayt. There he stayed with his slight wink, slight nod, and slight grin. His hands remained by his side.

After Wyatt's fall, Dr. Flo insisted that they step outside for some fresh air, but Wyatt didn't want to waste any time. He wanted to see if they could solve the mystery that had persisted all year.

Danni, Wyatt, and Dr. Flo stood back from the map. Scattered over the entire watershed were red and blue stickers. Wyatt was puzzled.

"It doesn't really tell us anything," Danni said. "It looks pretty random."

Wyatt nodded, disappointed. He had hoped some magic answer or great insight would suddenly jump out at him.

"Oh, I disagree," Dr. Flo said. "Sometimes you have to look carefully to make the connections to understand. Look at your red stickers. What do they mean?"

"Poor water quality," Wyatt said. Danni nodded.

"Where are they?" Dr. Flo asked.

"You mean, besides everywhere? Honey Brook and Wilmington," Wyatt answered. "And West Chester. And a few other places."

Dr. Flo fired questions at the two cousins. "Why Honey Brook? Why Wilmington? What is it about the few others? Think, think, think," she said.

Danni ran her finger across a line of red stickers from the Exton Mall through Downingtown to Coatesville.

"Connect the dots," Dr. Flo said.

And Wyatt saw it. It wasn't exactly an epiphany, but he saw it. "The poor water quality runs along that big road." He looked closely at the map to see the name. "Route 30. It's where the bigger towns are, like Downingtown and Coatesville. And that's where a lot of mills were."

"And what about up there?" Dr. Flo said, pointing to the northwest corner.

"That's Honey Brook, where there was all that open farmland surrounding the creek. And West Chester," Wyatt said, pointing to the southeast. "It's a big town, too. And then Wilmington," he traced a line toward the bottom of the map, "gets all that gunk added up from everything upstream." Suddenly, the models and demonstrations that he had seen over the last months made sense with the results plotted all over the map. The macroinvertebrates in the creek were a reflection of how the land affected the water they lived in.

"Look at the blue stickers," Dr. Flo said. "Do you remember any of those areas?"

Danni jumped in. "The good streams are in the woods, like at the Saalbach Farm and at Springton Manor and Hibernia Parks."

Map It!

"Or just downstream from wetlands," added Wyatt. "And the poor streams are where it's all open, along main roads, or where there are farms or lots of buildings." He thought back to the demonstration Dr. Flo and he had done months earlier, when he poured chocolate syrup, salt, powdered drink mix, soap, and oil into the container of water to form a mixture of slop. "It's really true. What you do to the land affects the quality of the water."

"It's true, yes," Dr. Flo said. "It always has been and it always will be."

Wyatt and Danni shot a look at each other.

"What's that supposed to mean?" said Wyatt.

~ ~ ~ ~ ~

~~Gladness~~
Sadness
Badness
Who will help to stop the Madness?
Y
ch44

Wyatt was restless. He tossed and turned in bed, unable to sleep. *Could this be what the woodblock referred to? Were the sadness, badness, and madness all due to pollution in the watershed? Something just didn't add up,* he thought. *And where did the message come from in the first place? And when?*

Chapter 27

TENNIS, ANYONE?

The drive to Honey Brook was long, but it was a sunny morning, and the June air that blew through the car window was energizing. Wyatt had to hold down Benny's flapping ears to see out the window. From long stretches of open farmland, to large housing developments, strip malls, towns, railways, industries, and forests, the landscape kept changing from one use to another, each having its own effect on the one thing they had in common: they all drained into the Brandywine Creek. Wyatt turned the woodblock in his hands. He had grabbed it when they left the house that morning, hopeful that this trip would have something to do with solving the mystery. Danni, who sat in the back of the car, tossed the white tennis ball from one hand to the other, like a metronome.

Map It!

A FOR SALE sign greeted them at the end of a long, gravel driveway that led to an old farmhouse. "I haven't been here in ages," Dr. Flo said. "Uncle Clayt used to take me on some of his trips around the watershed and, occasionally, we visited his friend Amos. Abe and I were your ages," she said, looking at Wyatt and then Danni. "We played together while the adults talked. We visited the pigs, groomed the horses, played in the creek, jumped in the hay bales in the barn. Those were the days," she sighed as she parked the car.

Abe emerged from behind the screen door of the farmhouse, a cup of coffee in one hand. Benny jumped from the car and greeted him, then began sniffing frantically around the property.

"I'm going to miss the old place," Abe said, handing the cup to Dr. Flo. She stood on her tippy toes to put an arm around his broad shoulders. "We're waiting for the township to finalize their offer to turn it into a park. In the meantime, we really have to keep the developers at bay. Wouldn't that be a great tribute to my dad? *Amos Stoltzfus Park*. And they could keep the house for a visitors' center or something."

"That's such a terrific alternative to a development," Dr. Flo said. Wyatt knew there was a lot of money to be made on a big piece of land. The pressure was great to sell to a developer, who could slice it up to build on. His father talked about it once in a while, since more buildings always meant more wastewater. On the other hand, parks were needed not just for play, but for maintaining a healthy environment and keeping the creek water clean. Living in Wilmington, Wyatt knew that the impact of living downstream of just about everybody and his land use had its consequences, but they didn't all have to be negative.

"Want to take a look around?" Abe asked. "Could be the last time you get to."

They entered the house. Sunlight streamed in through the east-facing windows, highlighting dust specks that floated in the abandoned space. A few pieces of furniture remained, but most of the house was empty, void of the life that once filled it.

"I can practically smell my mother's apple pie baking," said Abe.

"I remember your sisters playing dolls while your mother canned tomatoes," Dr. Flo said. "I think they always thought I was strange because I'd rather go explore with you than play with your sisters."

"You *were* strange," said Abe. "Correction, you *are* strange." Wyatt laughed, though he was anxious to get the chitchat over with and get on with the purpose of the visit.

The steps to the second floor squeaked as Benny ran up and the group followed. Abe pulled down the foldaway stairs to the attic and they climbed up. The attic was hot and dry and the air was stale. Abe walked over to an old wooden trunk that resembled a treasure chest. He opened the lid slowly. Benny cocked his head when the lid creaked.

"Got to oil that thing," Abe said.

A few items were scattered on the bottom of the chest: a rag doll, a scrapbook, a pair of framed needlepoint pictures. Benny fixated on something in the trunk. He barked.

"Hush now," Wyatt said. "Sit." Benny obeyed, but whimpered.

Abe lifted a torn paper sack from the trunk. "I thought this was trash. Until I found this." He pulled out a letter, but Wyatt couldn't stop staring at the sack. It was old and faded, but still readable: *The Draper Maynard Company, Tennis Balls approved by the United States Lawn Tennis Association. 100% wool felt covered; made in the USA;*

championship tennis balls. He asked Danni for the tennis ball they had brought.

"I left it in the car," she said.

"Well, go get it!"

"Funny," Abe said, while they waited for Danni to return. "You don't see this kind of thing anymore. They packaged tennis balls in paper sacks for just a few years during the war."

"Why?" asked Wyatt.

"Well, they saved metal for the war effort, to make weapons and tanks. They couldn't waste it on tennis ball containers, now, could they? That was the big one, World War Two, when the whole free world had to pitch in to stop Hitler."

"When *was* World War Two?" Wyatt asked.

Dr. Flo and Abe exchanged looks. "Don't they teach you anything in school?" Dr. Flo said. "Pearl Harbor was bombed in December 1941, and the war lasted until August 1945. Why?"

Wyatt didn't answer, distracted by Danni as she clamored back up the steps, the tennis ball in her hand. "Here it is." She tossed it to Wyatt.

"What's going on?" said Abe, scratching his head.

Wyatt stared at the ball. There it was, just as he had thought. *100% wool; USLTA.*

~ ~ ~ ~ ~

"That sack seems pretty important to you," Abe said, once they were all seated outside on the porch steps.

Wyatt explained. "We've been trying to figure out when this tennis ball came from. All we knew before was that it was from sometime between 1920 and 1975, when tennis was called *USLTA*. If the ball came from a sack like this, it would really help us solve a mystery."

Abe still held the letter. "This is what I *thought* was important." He handed it to Dr. Flo, who looked it over.

"I think *you* should have this," she said and passed it to Wyatt.

He handled it carefully. It was yellowed with age and fragile. Danni moved closer to peer around his shoulder. She stroked Benny as he lay calmly next to her. Wyatt took a deep breath and read the letter aloud.

"To Someone Who Cares:

I write this letter with a heavy heart as I rest at the home of my good friend, Amos Stoltzfus. My dog, Max, no longer lies beside me as he did for fourteen years, resting in preparation for our daily walk to the creek for a game of fetch. There's nothing Max liked more than to splash into the sweet waters of the Brandywine to retrieve a tennis ball, even on a chilly, cloudy November day such as this. Max passed away yesterday here up in Honey Brook, and we buried him in a wooded area where the trees will shelter him from the cold winds of winter and shade him from the heat of summer.

Max had a good and full life, so I cannot attribute my heavy heart to his death alone. I have traveled this valley for years with Max, from Wilmington to Honey Brook. And I have become disturbed by the sights and smells of the Brandywine that was once so magnificent. After all, how else could an area have attracted and sustained so many different people from the Lenape and the Quakers, to the Amish, from the Swedes and the Scots, to the Welsh, from the millers and the industrialists, to the socialites? Where else in this fledgling country would William Penn want to settle other than the fertile, forested, rich lands of the Brandywine?

But the settlements took their toll and the sweet waters that once thrived with shad to feed the Lenape and powered the mills of past centuries have become an abomination. Even the name Brandywine has an inauspicious folklore: did it come from the early settler and miller Brainwende or from the early pollution of a Dutch wagon that crashed into the creek and dumped its load of brandy? There is a disregard for our natural world, and the Brandywine Creek has become a consequence, as well as a warning. In too many places, it has become an open sewer, a recipient of the short-term, ill-advised land uses that surround it. This creek of such vast and important history in the formation of our great nation has been taken for granted and left to fend for itself against the damage we inflict with our sewage, our runoff, our chemicals and our waste.

Today, I grapple with the loss of my beloved Max, an inevitable arc in the circle of life. But I grieve as well for the impairment of the Brandywine, an avoidable, but possibly redeemable, condition.

I will remain here at the Stoltzfus farm for several days before I journey home to Wilmington. I will sort through my thoughts and concentrate not on the irreversible state of Max's death. For that I can only grieve. But I will focus on the future, on the possibility of redemption for this valley and its waters, for a way to stop the thoughtless treatment of our environment, for a way to stop the ignorance of the average person who contributes to this state of affairs. I wish myself the best of luck.

Signed,
Clayton Hoff
November 29, 1944"

"I remember Max," Dr. Flo said, breaking the silence. "He was a great dog. And I remember how he whined when Uncle Clayt sometimes decided against playing fetch in the water for fear of making him sick."

Wyatt thought back to the day Benny retrieved the tennis ball last summer at the park, and the signs warning people not to go into the water because of the pollution. *But Clayt Hoff's letter was written more than sixty-five years ago,* he thought. *Has nothing changed?*

Danni held up the tennis ball. "Does this have something to do with the letter?" she asked, confused. Benny looked up, alert and hopeful that Danni would throw the ball for him to fetch.

Wyatt pulled out the woodblock and read aloud.

~~Gladness~~
Sadness
Badness
Who will help to stop the Madness?
Y
ch44

"Don't you get, Danni?" Wyatt said. "*ch44* is Clayton Hoff, 1944. *He* carved the woodblock and it was *his* tennis ball."

Chapter 28

THE VIEW

"In March of 1945, thirty people from the West Chester and Wilmington areas got together to listen to a man named Clayton Hoff talk about the Brandywine Creek. What they heard was alarming. In many spots the creek was little more than an open sewer, the result of wastewater being dumped into the stream with little or no treatment. In addition, thousands of tons of soil were being washed into the Brandywine—choking aquatic life and diminishing water quality. Recognizing that such threats would cause permanent damage, this group founded The Brandywine Valley Association, the first small watershed association in America." —www.brandywinewatershed.org

Just a few minutes after leaving the farmhouse, Dr. Flo's car chugged its way up the long, winding slope of the Lanchester Landfill. Small mountains filled with trash surrounded them. Some were capped permanently, covered with grass. Others were still in use and covered with just a layer of soil to prevent the trash from blowing around overnight. Overhead, gulls soared. A piece of paper drifted by.

Check It Out!

When they arrived at the landfill, Wyatt was a little surprised. A landfill seemed like a strange place to go for a picnic lunch, even with Dr. Flo. But it was close by to the Stoltzfus farmhouse and they were all hungry.

Map It!

They drove up the man-made mountain until they reached a flat vista where there was a pavilion, a small playground, and a continuously spinning windmill. Dr. Flo parked the car. They stepped out and were surprised by how windy it was. Benny strained on his leash. *I bet there's some serious smelling to do here*, thought Wyatt. Since there were no other people in sight, he let Benny off the leash.

"Can you believe such a beautiful view could come from all of our trash?" said Dr. Flo. Wyatt and Danni turned around to take in the scenery. In all directions, east, south, west, and north, the view of valleys and rolling hills was spectacular. The patchwork of yellow and green fields, forests, and developments looked like a quilt from the Chester County Historical Society.

"To the north and west," Dr. Flo pointed out, "is the Conestoga Valley; all that land drains into the Conestoga River. Do you know what you're looking at to the south and east?"

"Well, I hope that's the Brandywine Watershed," Wyatt said.

"What you are looking at is what you have been exploring this past year. The sum of all of its parts. The Brandywine Valley."

Danni thought back to the Revolutionary War actor. *E pluribus unum,* she remembered him saying. *The many that become one.*

Wyatt stayed quiet.

"What is it?" Dr. Flo asked.

Wyatt tried to sort out the questions swirling in his head. "Well, for one thing, look at how nice everything looks from here. It doesn't look polluted. And the creek doesn't look like an open sewer, like your Uncle Clayt said it was." He paused. "And how come we just found the tennis ball? Didn't he carve the woodblock, like, sixty-five years ago?"

"Come, let's sit down," Dr. Flo said, and they headed over to a picnic table overlooking the Brandywine Valley. ELEVATION: 1100 FT was etched into a corner of the table.

"You are looking at the three hundred and thirty square miles of the Brandywine Watershed," Dr. Flo began. "Three hundred and thirty square miles of woodlands and wetlands, communities and waterways, industries and parks. From the Welsh Mountains up here in Honey Brook, all the way to Wilmington, where we live, the Brandywine Creek travels sixty miles through two different states, four counties, and forty-four municipalities. It doesn't recognize political boundaries or understand the various cultures that have impacted it. It just flows, cycling water from the ground through our communities and eventually out to the ocean. My Uncle Clayt realized, through his travels, just how important the Brandywine Creek is for our very existence. And he realized how much damage we have done to it through the centuries.

By clearing the forests and damming up the waters for water power, the colonists changed the course of the waters, forcing the Lenape Indians to abandon their livelihood. The rise of industries led to industrial wastes being dumped into the creek. Once that was addressed, the pollution from runoff from our farms and our developments became the problem. And just recently, we've identified legacy sediments, built up behind old milldams, as another culprit of sediment pollution.

"Each generation, it seems, has contributed to the degradation of the Brandywine, yet the creek still survives and continues to provide us with water, wildlife, and recreation. When Uncle Clayt wrote his note back in the forties, there were some serious problems in the Brandywine. Many of those problems have been addressed. Now, there are different problems that require different solutions."

Wyatt thought for a moment before speaking. "So the madness that your Uncle Clayt was writing about isn't the same madness that there is today?"

"I'm sure it's not. Uncle Clayt worked to stop the madness he saw back then by rallying people together to understand and do something about that pollution. He's known as the founder of the small watershed movement, you know," Dr. Flo said proudly. "'*Never doubt that a small group of thoughtful, committed citizens can change the world. Indeed, it is the only thing that ever has.*' Do you know who said that?"

"No idea," said Wyatt. Danni shook her head.

"Margaret Mead, a famous anthropologist."

"Was your Uncle Clayt famous?"

"Not really," Dr. Flo said. "Most people take water for granted. They turn on the tap and expect clean, fresh water to come out. They flush the toilet and expect their wastes to disappear. They don't realize that the Brandywine is, today, forty percent impaired. They don't realize what it takes to keep water clean and flowing. They don't realize how they sometimes pollute it themselves. So they certainly don't think about the people who keep it safe for them. Like Uncle Clayt did. And like your father does now, Wyatt."

Wyatt thought for a moment about his father and his work. Even if it was a gross job, it was a really important one. "So, if we know about all the ways that water gets polluted, is there still madness now?"

"What do you think?" Dr. Flo answered his question with a question.

~ ~ ~ ~ ~

Wyatt slept peacefully and dreamt deeply.

He climbed into the canoe, carefully, so as not to tip it. Jennie turned and smiled at him. "Remember last time?" she giggled. He shoved off and started to paddle. Downstream, along the shoreline, a small group of people was gathered. As the canoe approached, the people turned toward him. Each one looked familiar. As he passed by, each of them, one by one, gave a thumbs up, winking and smiling: the du Pont boy; Uncle Clayt; the disheveled man from Hagley; Indian Hannah with her pigs and dogs; Deborah; Horatio Myrick; Rebecca Lukens. "Look, Wyatt," said Jennie. "They must really like you." Four crows flew overhead.

Field Trip!

Chapter 29

SUMMER SOLSTICE

The annual Nystrom picnic dinner at Brandywine Park took place on the longest day of the year. After feasting on fried chicken, deviled eggs, and homemade brownies, the families would linger in the park until dusk, which didn't come until nearly eight-thirty.

Meg and June gathered the trash from their dinner, while Luke and John dozed on a blanket.

"C'mon," Wyatt called to Danni. They led Benny toward the bridge. When the dog realized where they were going, he darted ahead, tugging at the leash. They crossed over the creek and scrambled down the steps on the far side of the bridge.

"Yuck," said Danni, pointing at a pile of trash built up behind some rocks in the creek. A beer can and a pair of swim trunks were distinguishable in the clutter. "Doesn't someone know when he loses his bathing suit?" she asked.

"Maybe he was skinny dipping," Wyatt answered. But he was thinking about something else. *How long will those swim trunks be there? Could they stay stuck there for years? For sixty-five years? Or maybe they've been making their way down the creek for decades, and have just recently arrived here. Maybe those trunks came from Honey Brook. Maybe they were Uncle Clayt's!*

Benny's barking broke Wyatt's concentration.

"All right, all right," he said. They walked to the bank of the creek and let Benny off the leash. "Look what I brought for you, boy." Wyatt reached into his pocket and pulled out the old tennis ball. "This one's for Uncle Clayt and Max," he said. He hurled the ball into the creek and Benny dashed after it.

Epilogue

Clayton Hoff sat on the porch steps of his friend's farmhouse on a day late in November. He didn't notice the chill in the air as he carved words into the small block of wood he found in the workshop. Gouging the wood helped distract him from his grief over the death of his beloved dog, Max. Word by word, he carved the sides of the woodblock. Then he carved his initials and the year: 1944.

He kissed Benny's white tennis ball, sliced it cross-wise, and pushed the woodblock in. "I'm going to take a walk," he called to Amos's wife.

"Don't you go anywhere without this," she scolded. She took Amos's cloak from its hook by the kitchen door and handed it to him.

Clayt shrugged on the cloak and left the farmhouse. He hiked across the field toward the old sycamore tree. The sky had thickened with dark clouds and the wind had picked up. Overhead, four black crows flew and then landed on several tree limbs above him. The autumn leaves swirled around as he knelt beside the creek and released Max's tennis ball for what he thought was the last time.

Appendix A

ACTIVITIES TO DO WHILE YOU READ *SWEET WATER HUNT*

As you read the novel, you will be prompted to **Map It!, Check It Out!, Sticker It!, Try It!** or go on a **Field Trip!** An explanation of what to do for each of these suggested activities follows. In addition, throughout the novel, you can create a **FIELD GUIDE**. Though not prompted to do so throughout the text, you will read about a variety of plants and animals encountered by the characters. You can create an ongoing field guide of the biology of the Brandywine Valley. Each time that an organism is mentioned, you can research its natural history, draw it, and create a field guide entry. An example of such an entry can be found in Appendix P. For more information on activities, events, teacher workshops, and programs related to Sweet Water Hunt, please visit www.sweetwatered.com

Map It!: Use your Brandywine Watershed map to identify sites referred to in the text. The numbers on the map correlate to the chapters. You may want to flag the site or simply identify the area so that you understand the context of the location as the story progresses.

You can use the map included with the novel, or, to be able to map out the sites more easily, the author suggests investing in a set of maps that can be taped together, as Dr. Flo does. This can be done with USGS topographical maps (recommended) or with street maps like those found in the ADC Street Map books (you will need the Pennsylvania map books for Chester County, Delaware County, Lancaster County, and the Delaware map book for New Castle County). When taped together, either set of maps will cover a space of about 6' wide by 10' tall. If using USGS topo maps, the following quadrangles are needed: Wilmington South, Wilmington North, Kennett Square, West Chester, Unionville,

Coatesville, Parkesburg, Malvern, Downingtown, Wagontown, Honey Brook, Elverson and Morgantown.

Check It Out!: This alerts you to do some research. Use your library, the Internet or other resources to check for background information. Often a website will be offered as a starting point, but go as far as your interest will take you. Be warned: websites come and go, so some in this appendix may not be available by the time you try to link to it. Just find another!

Sticker It!: This is when you place either a red sticker or a blue sticker on the map of the Brandywine Watershed at the site indicated in the text. These stickers indicate the quality of the water. For simplicity's sake, a red sticker indicates poor water quality (a Biotic Index of less than 10) and a blue sticker indicates good water quality (a Biotic Index of 10 or more). More about this will be explained in the novel. The more stickers you place on the map, the more you will be able to discern some patterns, though you may not be able to see "the big picture" until all the stickers are on the map and Dr. Flo, Wyatt, and Danni help you to recognize the patterns.

Try It!: This refers to an actual activity you can try. Sometimes the activity will be described in the text of the novel, with no further explanation needed. Sometimes a fuller set of directions will be given in this guide or in another appendix.

Field Trip!: You can actually go to the site for a field trip on your own or with your class. Check the links for more information. Please be aware that roads have changed since the novel was written (i.e. Route 52, where Indian Hannah's birthplace monument is, has been relocated), so you may need to do some research to find sites referred to in the text.

PART I: The Christina River to the Brandywine Confluence

Chapter 1: The Swedes Land (July)

Check It Out! Look into the history of Delaware's earliest governors.

Map It! Find where the Christina River joins the Delaware River, where the Delaware River meets the Delaware Bay, and where the Delaware Bay meets the Atlantic Ocean.

Check It Out! Research the Lenape Nation's *The Prophecy of the Fourth Crow,* and look up the meaning of crows for the Lenape Indians.

Map It! Find *The Rocks* in Wilmington, where the Kalmar Nyckel first landed. This is also labeled as *Fort Christina Park* or *Fort C Park.*

Check It Out! Research information about Queen Christina and her abdication of the Swedish throne.

Sticker It Red! Place a red sticker on the Christina River just upstream from where the Brandywine Creek merges.

Try It! Whenever you see lightning, count the number of seconds until you hear the thunder that accompanies it. Every five seconds indicates one mile of distance from the lightning. Since sound (thunder) travels more slowly than light (lightning), there is a delay, even though both the sound and the light originate from the same place at the same time.

Field Trip! You can book a trip on a replica of the Kalmar Nyckel ship. For information, go to http://www.kalmarnyckel.org/schooltrips.asp.

Chapter 2: The Hunt Begins (July)

Map It! and Sticker It Red! Locate Brandywine Park in Wilmington and place a red sticker in the Brandywine Creek at this location.

Try It! Try doing crayon or pencil rubbings of various outdoor textures. Try rubbing leaves or the bark from different trees. Be sure to do a rubbing on a sycamore tree!

Try It! Check your reflection in a round, metal door knob. Learn the difference between *convex* and *concave.*

Field Trip! Take a trip to Brandywine Park in Wilmington. Within walking distance is the Brandywine Zoo.

Chapter 3: The Riddle is Revealed (July)

Map It! Find Alapocas Woods Park along the Brandywine, just north of Wilmington.

Map It! Find where Alapocas Run joins the Brandywine Creek.

Check It Out! Check track prints of various animals of Southeastern Pennsylvania. Do an online search or use a tracking guidebook like the Audubon Society Pocket Guide to Familiar Animal Tracks of North America. Look online to make plaster casts of tracks.

Check It Out! Go outside, close your eyes and simply listen to the natural sounds around you. To identify specific animal sounds, do an online search or use a Birdsong Identiflyer (for information or a demonstration, visit www.identiflyer.com).

Check It Out! Research the life cycle of a mayfly insect.

Check It Out! Research planaria regeneration. You can even try this one!

Try It! Pick rocks from a stream bed and see if there are critters or homes on the underside. Using the key in Appendix C, try to identify what you find.

Check It Out! On a large-scale map that includes Southeastern Pennsylvania, trace a line from Alapocas Run to the Brandywine Creek to the Christina River to the Delaware River to the Delaware Bay to the Atlantic Ocean. Or try using Google Earth.

Field Trip! Take a trip to Alapocas Woods Park to see the climbing rocks and the old Indian settlement grounds. Find where Alapocas Run joins the Brandywine Creek.

Chapter 4: What's a Watershed (August)

Map It! Find Wawaset Park in Wilmington.

Map It! Find Riddle Avenue and Wyatt's house on Mill Road (along the Brandywine).

Map It! Find Highland Elementary School and Warner Elementary School (both in Wilmington) and Winterthur Museum and Gardens (just north of Wilmington).

Try It! Follow Dr. Flo's directions and try "Knuckle Mountain" on your own knuckles. Or visit http://www.education.com/activity/

article/knuckle-contour-map/ for a more complete activity description.

Try It! Around Winterthur, do as Wyatt does and highlight the streams around Wilson Run.

Try It! Mark each of the highpoints around Wilson Run with an "X" and connect the "Xs" to indicate the Wilson Run watershed.

Try It! Highlight Wilson Run to where it joins the Brandywine and then continue highlighting the path the water flows until you reach the Christina River, Delaware River, Delaware Bay and Atlantic Ocean.

Try It! Make a batch of Dr. Flo's cookies. See the recipe in Appendix B.

Check It Out! On a map of the United States, find the Continental Divide (mostly along the Rocky Mountains). Note the elevations of the CD; anything east of the CD is in the Atlantic Watershed; anything west is in the Pacific Watershed.

Check It Out! Research toilet-lid sinks. An informative website is: http://life.gaiam.com/article/water-conservation-tips-toilet-lid-sink-faq.

Check It Out! Research various insects that undergo metamorphosis and the different types of metamorphosis: complete and incomplete.

Try It! Try different math problems related to bats. Bat Conservation International's *Educator's Activity Book About Bats* has great activities, or go to http://mathwire.com/themes/themebat.html.

Try It! Follow Dr. Flo's directions and replicate the activity she and Wyatt do. Be sure to relate each ingredient to a likely real-world pollutant. As a guide, you can follow the directions in "The Story of Our Town" in Appendix E. OR contact a local water protection organization (Chester County Conservation District or Brandywine Valley Association) and borrow an "Enviroscape" model to demonstrate point and nonpoint source pollution. Save the polluted water for an activity in Chapter 16 (or redo this pollution activity).

Field Trip! Take a field trip to a pond and do a study to understand Dr. Flo's bathroom art.

Chapter 6: Fall Field Trip (September)

Map It! Find the Hagley Museum on the Brandywine Creek, just north of the city of Wilmington and just south of Winterthur Museum and Gardens.

Check It Out! Research the history of making gunpowder and its various uses.

Check It Out! (A brief note about the du Ponts: Various spellings of the name are found in various places and among various members of the family. For the sake of consistency, the author has, generally, chosen to use the du Pont spelling for family members and the DuPont business spelling for all other uses of the name throughout the manuscript.) Research the DuPont Company and its history. It all started at Hagley! Identify 5-10 products that DuPont makes.

Field Trip! Take a field trip to the Hagley Museum. For more information, visit http://www.hagley.lib.de.us/info.html.

Chapter 7: Creek Pals (September)

Map It! If you didn't already do this in Chapter 4, find Warner Elementary in Wilmington, Delaware.

Map It! Find Blackshire Road in the Wawaset Park neighborhood just south of Highland Elementary School.

Check It Out! Research the names of animal groups, including crows.

Check It Out! Research the origins of Halloween customs, as well as the various meanings of witches and ghosts.

Try It! Collect leaves and try classifying and/or identifying them. A simple, free leaf-identification key for Pennsylvania trees can be found at: www.downloads.cas.psu.edu/4H/Summerkeyfortrees.pdf.

Check It Out! Listen to some of the oldies of Dr. Flo's. Be sure to listen to *Witch Doctor*, a familiar tune to everyone.

Map It! Locate Chadds Ford Elementary School and Ring Run, which flows into the Brandywine near the elementary school.

Check It Out! See Appendix C for the Biotic Index Chart and Procedures for conducting a water quality test.

Chapter 8: Brandywine Beauty Mark (October)

Map It! Locate Brandywine Creek State Park (it's different from Brandywine Park in Wilmington). It is located on the Brandywine Creek, north of Wilmington, but south of the PA/DE Border.

Try It! Learn to measure temperature, turbidity, width, depth and velocity of a creek. See Appendix D.

Check It Out! Research the meaning of pH and its significance to the biology of a creek. See Appendix D.

Check It Out! Research the meaning of dissolved oxygen and its significance to the biology of a creek. See Appendix D.

Check It Out! For photos of macroinvertebrates, use Appendix C or visit www.stroudcenter.org/research/projects/schuylkill/macroslideshow.shtm#taxa/images/taxon0. Or you can order stream macroinvertebrate cards through the Lamotte Company: http://www.lamotte.com/pages/edu/5882.html.

Check It Out! Research various ways to determine water quality based on the biology of the stream. Check www.saveourstreams.org or check with your local watershed association. See Appendix C for Dr. Flo's standard test for younger students.

Field Trip! Visit and hike Brandywine Creek State Park. Go to a creek and run the physical, chemical, and biological tests as described on the instruction sheet in Appendixes C & D. Record all data on the data sheets provided in Appendixes C & D. Or go to a nature center for a stream-study program that includes physical, chemical, and biological testing.

Chapter 9: Dream On! (October)

Sticker It Red! Place a red sticker on the Brandywine at Brandywine Creek State Park.

Map It! Locate the Brandywine Creek at the Pennsylvania/Delaware border.

Check It Out! For guidelines to permits for collecting macroinvertebrates, go to www.fishandboat.com/education/collinfo.htm.

Check It Out! Do a search for explosions at DuPont powder mills and see what you find.

Check It Out! Research the genealogy of the du Pont family and some of its important family members.

Chapter 10: Getting to the Art of It (October)

Map It! Locate the Brandywine River Museum in Chadds Ford on the Brandywine Creek.

Rap It! Turn the poem into a rap song. See Appendix F for the entire song.

Check It Out! Check out the scat of various animals, including fox. For help, visit http://www.terrierman.com/scatanswers.htm. To test yourself on scat identification, take the "Animal Scat" quiz at http://www.northwoodsguides.com.

Check It Out! Research the plants, animals and functions of a freshwater wetland. Or observe a wetland.

Check It Out! Research the artwork of the various Wyeth family artists. Read *Pieces of Georgia* by Jen Bryant about the Brandywine River Museum.

Check It Out! Research the benefits of wetlands. Or go to a nature center for a wetlands program.

Map It! Locate Hillendale Elementary School, just east of Longwood Gardens

Map It! Locate the intersection of Woodchuck Way, Turkey Hollow Road (just a little northwest of Hillendale Elementary School), and Ring Run, the creek that runs through it.

Map It! Trace the creeks around Eliza's house to either the Brandywine Creek or the Red Clay Creek.

Map It! Locate Longwood Gardens along Route 1, northeast of Kennett Square.

Check It Out! Research some of the du Pont family members and the contributions they made to the area.

Map It! Locate the area that used to be Route 52 between Routes 1 & 926, just north of Longwood Gardens, where the Indian Hannah marker is (location: 39°52.253'N; 75°40.101'W).

Check It Out! Research and learn about how the Lenape Indians lived. Or go to a nature center for a Lenape Indian program.

Map It! Locate the boundary between Pennsylvania and Delaware on Route 100.

Field Trip! Visit the Brandywine River Museum, Longwood Gardens or a nature center with a wetlands program like Brandywine Valley Association.

Chapter 11: Clear Skies (October)

Check It Out! Do the math. If it took the tennis ball six days to travel three miles, how long would it take the ball to travel 60 miles?

Sticker It Blue! Place a blue sticker on Ring Run by Chadds Ford Elementary School.

Sticker It Red! Place a red sticker on the Brandywine Creek in Chadds Ford, just north of the Route 1 bridge.

Map It! Find the area just northwest of the intersection between Routes 1 and 100, where the Great Pumpkin Carve takes place.

Check It Out! Research various Lenape words and phrases. Go to the Lenape talking dictionary at www.talk-lenape.org.

Field Trip! Visit Chadds Ford annual "Great Pumpkin Carve" at the end of October.

Chapter 12: Marching Orders (November)

Map It! Locate the setup for the Battle of the Brandywine at Chadds Ford and Kennett Square.

Check It Out! Research and create a timeline of the events of the Revolutionary War leading up to the Battle of the Brandywine and the events following this Battle to the Ratification of the Constitution. See Appendix H for Mrs. Bibbo's example.

Map It! Find the areas where the Patriots set up their troops along the Brandywine.

Map It! On a map of the eastern U.S., locate the route the British took from New York to the Chesapeake Bay and locate Kennett Square.

Check It Out! Listen to the theme song from the movie "Rocky."

Map It! Find Trimble's and Jefferis' Fords along the Brandywine.

Map It! Locate Birmingham Friends Meeting House, now a historical site just east of the intersection of Routes 926 and 100 in Birmingham Township.

Try It! After researching the Battle of the Brandywine, try reenacting it with toy soldiers on a map, like Danni's class.

Check It Out! Visit www.nedhector.com for information about this Revolutionary War character or to book a school visit.

Check It Out! Research various Lenape words and phrases. Go to the Lenape talking dictionary at www.talk-lenape.org.

Map It! Locate Brandywine Battlefield Park on Route 1 just east of Chadds Ford. Note that the park is not where the actual battle took place.

Check It Out! Learn about the Brandywine Valley Association's Red Streams Blue program. Visit http://www.brandywinewatershed. org/2008/redstreamsblue/index.asp.

Field Trip! Take a trip to the Brandywine Battlefield and Visitor's Center. For more information, visit http://www. brandywinebattlefield.org/.

Chapter 13: Quest (November)

Map It! Locate Pocopson Elementary School between Routes 52 and 1 on the Brandywine Creek.

Map It! Locate Pocopson Home on Route 52, a little west of Pocopson Elementary School.

Map It! Locate Chester County Prison.

Map It! Locate the site of the Brandywine Valley Association/Red Clay Valley Association/Myrick Conservation Center. Though it is not designated on the map, it is located on Route 842, just east of Northbrook and Red Lion Roads.

Check It Out! Read *Notes from Turtle Creek*, a compilation of Ted Browning's essays written for *The Kennett Paper*. The book was published after his death.

Check It Out! Research composting toilets and their advantages.

Rap It! You know what to do! See Appendix F.

Check It Out! Use Appendix I to trace the community water cycle, following water from its source (like the Brandywine Creek) to the home and back to the source.

Check It Out! Research the history of Velcro. Go outdoors and examine a burr and compare it to Velcro.

Field Trip! Visit the BVA/RCVA/MCC and go on the Quest. Or schedule one or two of the numerous outdoor, hands-on classes offered to school groups. For information, visit: www. brandywinewatershed.org.

Chapter 14: Go With the Flow (November)

Map It! On a map of the Brandywine Creek, highlight the East and West Branches and the Main Stem of the creek to visualize the "Y" shape.

Try It! Reenact the Brandywine by following Dr. Flo's directions in the text.

Map It! Highlight all the feeder streams to the West Branch of the Brandywine in one color, and all of the feeder streams to the East Branch of the Brandywine in another color.

Try It! Find a highpoint between the East and West Branches of the Brandywine in Honey Brook and mark it with an "X." Note the elevation. Find an elevation close to Wilmington and compare it to the one in Honey Brook. Determine which way water flows.

Sticker It Red or Blue! Place a red or a blue sticker on each site described in the chart.

PART II: Above the Confluence

Chapter 15: Rock Legend (December)

Map It! Locate Natural Lands Trust's Stroud Preserve. It is on the East Branch of the Brandywine just south of Route 162.

Check It Out! Look up kingfishers and red-tailed hawks. Use the Birdsong Identiflyer or other tools (online or "talking books") to listen to their calls as well as the calls of various other local birds.

Rap It! See Appendix F and you know what to do!

Map It! Find Taylor Run. Trace it upstream to the wastewater treatment plant.

Check It Out! Check out the scat of various animals, including coyote. For help, visit http://www.terrierman.com/scatanswers.

htm. If you didn't do this already in Chapter 10, test yourself on scat identification at http://www.northwoodsguides.com.

Check It Out! Check into the Marquis de Lafayette's role in American history and Abraham Lincoln's ties with Chester County.

Check It Out! Research the controversy of quilts used as part of the Underground Railroad.

Check It Out! Research the basics of the Underground Railroad and the line between the slave and free states. Check out the map with clickable Underground Railroad sites in the Brandywine area at www.undergroundrr.kennett.net. Also, the Chester County Historical Society (www.chestercohistorical.org) has a nice brochure called "Just Over the Line" as well as sets of story cards about important local characters from the Underground Railroad.

Check It Out! Read the picture book *Rainbow Crow* by Nancy Van Laan.

Check It Out! Read the picture book *How the Stars Fell Into the Sky* by Jerrie Oughton.

Check It Out! Research and learn about how the Lenape Indians lived. Or go to a nature center for a Lenape Indian program.

Field Trip! Visit Natural Land Trust's Stroud Preserve, the Chester County Historical Society in West Chester and/or West Chester's Old Fashioned Christmas (first weekend in December).

Chapter 16: Royal Flush (December)

Try It! Do the celery demonstration as Luke Nystrom describes. Check it after a few hours and a few days. Relate it to plants in riparian buffers and/or wetlands that absorb pollutants.

Map It! From Taylor Run, trace the course treated wastewater takes to the Atlantic Ocean.

Check It Out! Visit and find out how your local wastewater treatment plant cleans water. To demonstrate some of the filtration processes, use the polluted water you created in Chapter 4 (or make a new solution of polluted water) to do "Dirty Deeds" in Appendix G.

Sticker It Red or Blue! Place a red or blue sticker on the sites described in the chart.

Try It! Replicate the activity that demonstrates dilution.

Try It! Create and demonstrate the effectiveness of a soil filter. See Appendix J for instructions. Or visit a nature center with a soil filter for a demonstration.

Try It! Do the activity Runoff Race from *WOW! The Wonders of Wetlands;* see Appendix K for abbreviated instructions.

Check It Out! Research storm drains and storm drain stenciling projects. Check out http://www.protectingwater.com/storm-drains.html for a quick look at stormwater pollution.

Field Trip! Tour a wastewater treatment plant in your local community. In the Brandywine Watershed, there are plants in West Chester, Downingtown, Coatesville and Wilmington.

Chapter 17: Longest Night of the Year (December)

Try It! Make an animal edible treat by stringing popcorn and fruit pieces on a thread. Hang it outside on a tree for the animals.

Map It! Locate the course that Wyatt's family takes on this journey.

Map It! Locate and follow railroad tracks that connect Wilmington to Coatesville.

Check It Out! Research the Mason-Dixon line and the Fugitive Slave Act.

Map It! Locate Stargazer's Rock on Stargazer's Road in Embreeville.

Map It! From Route 162 in Marshalton, follow Marshalton-Thorndale Road. Note how it is a watershed boundary between the East and West Branches of the Brandywine. Locate where the East and West Branches of the Brandywine merge.

Map It! Locate the historic district of Marshalton.

Map It! Locate the historic marker of Indian Hannah's burial on the property of Embreeville Center/State Police Barracks.

Map It! Locate the ChesLen Preserve in Newlin Township, south of Route 162 around Cannery Road.

Check It Out! To learn more about local Lenape history, visit www.Lenapenation.org.

Try It! Try saying this Lenape word. For help with pronunciation, visit http://www.talk-lenape.org/index.php and search for the pronunciation of the English word "peace."

Check It Out! On a clear night, observe various constellations and the Milky Way. A great reference for constellation viewing is H.A. Rey's *The Stars-A New Way to See Them*.

Map It! Locate the marker for this Indian burial ground on Brandywine Drive between Warpath and Northbrook Roads. The historical marker is blue with yellow lettering.

Map It! Locate the marker for Indian Rock, just downstream from the Indian burial ground marker.

Try It! Visit http://www.talk-lenape.org/index.php and search for the word wanishi.

Field Trip! There are sites to visit in this chapter, including Stargazer's Rock, Highland Orchard, the historic district of Marshalton, Indian Hannah's burial site, and the ChesLen Preserve (including Potter's Field). Also look for the historical markers for the Indian Burial Grounds and Indian Rock along Brandywine Drive close to the Northbrook bridge.

Chapter 18: Christmas Day (December)

Map It! Locate Modena on the West Branch of the Brandywine, south of Coatesville.

Map It! Locate East Fallowfield Elementary School, southwest of Modena, and Ercildoun, west of EFES on Route 82.

Check It Out! Research Ida Ella Ruth Jones, an African American artist from this area. Look for the blue-and-yellow historical marker on Route 82 in East Fallowfield Township.

Check It Out! Research the history and sport of fox hunting.

Map It! Locate and research the history and significance of the King Ranch in Chester County along Route 82, south of Ercildoun and north of Unionville. You can find information at www. brandywineconservancy.org.

Check It Out! Research the history of the tennis ball.

Field Trip! Take a ride down Route 82 to observe the King Ranch conservation area. If you are a member of the Brandywine Conservancy, hike into The Laurels Preserve to find McCorkle's Rocks.

Chapter 19: Nature at Night (January)

Map It! Locate Paradise Farm Camps, on Valley Creek Road just south of Downingtown.

Try It! Using a flashlight and a mirror in a dark space, observe how your pupils change with the light on and the light low. Or you can do this by observing a partner instead of a mirror.

Map It! Locate the schools: Warner Elementary, Mary C. Howse Elementary, Church Farm School, Collegium Charter School, and Exton Elementary.

Try It! Dissect owl pellets and imitate the calls of various owls of the area including Great Horned Owls, Screech Owls and Barred Owls. Use a Birdsong Identiflyer (for information or a demonstration, visit www.identiflyer.com).

Check It Out! Research the symbolism of owls throughout history and in various cultures. Research the different types of owls in the Brandywine Valley.

Try It! Try various night sky activities. For a variety of astronomy activities, refer to Ranger Rick's NatureScope: *Astronomy Adventures*. And don't forget to have a campfire and make and eat S'mores!

Check It Out! On a clear night, go outside and do some astronomy observations. A great reference for constellation viewing is H.A. Rey's *The Stars-A New Way to See Them*.

Check It Out! Research what a light year is and how close and far some astronomical objects are. Do light-year math. Try "Take a Look at Light Years," from Ranger Rick's NatureScope: *Astronomy Adventures* or visit online light year math sites like http://school. discoveryeducation.com/schooladventures/universe/itsawesome/lightyears/index.html.

Field Trip! Schedule a field trip to Paradise Farm Camps. Visit www. paradisefarmcamps.org for a listing of school programs.

Chapter 20: Snow Day (February)

Map It! Locate East Ward Elementary School on local Route 30 in Downingtown.

Map It! Locate the old log cabin on Route 30 at Kerr Park in Downingtown.

Check It Out! Research daily life in America in the early 1700s and the 1800s.

Map It! Locate Shamona Creek where it joins the East Branch of the Brandywine, about 1.5 miles north of the center of Downingtown.

Check It Out! Explore, hike, or bike the full length of the Struble Trail and/or the Uwchlan Trail. On the Uwchlan Trail, follow and read the signs describing the forge along the Shamona Creek.

Check It Out! Research floods in Downingtown, including those caused by Hurricanes Agnes and Floyd.

Check It Out! Research, list, or illustrate the things we need fresh, potable water for. You can also try to find out what Downingtown's "Amphibious Order of Frogs Dinner" is. Good luck!

Map It! Locate Beaver Creek and Beaver Creek Elementary School.

Try It! Try the "Mummy Wrap" activity (using cloth strips) as Principal Lawless does to demonstrate the percentage of water in the human body (about 75%)

Try It! Try the activity "Limits of Water" in Appendix L.

Check It Out! Demonstrate and review the water cycle. Also, try the population growth activity found at http://www.worldof7billion. org/images/uploads/w7b_Population_Circle.pdf.

Map It! Locate the spot where Beaver Creek, Route 322 and Race Street come together in Downingtown. This is the location of Tabas Park and the Roger Hunt Mill.

Map It! Locate the Downingtown Municipal Water Authority (on Water Plant Way just north of Downingtown off of Route 282) and the Downingtown Regional Water Pollution Control Center (about 1 mile downstream of DMWA).

Map It! Locate where water is taken from the Brandywine (intake) for West Chester's and Wilmington's water treatment plants. West Chester's intake is located about two miles south of Downingtown's wastewater treatment plant on the East Branch Brandywine. The water intakes for Wilmington's two water treatment plants are in the Brandywine Park area, near the

Brandywine Zoo, about 15 miles south of Downingtown's wastewater treatment plant.

Field Trip! Visit the Downingtown Municipal Water Authority and the Downingtown Regional Water Pollution Control Center for tours. On the way, drive by the other sites mentioned in the chapter, like Tabas Park and the Roger Hunt Mill.

Chapter 21: Taking the Plunge (February)

Field Trip! See the Polar Plunge at Brandywine Picnic Park. The annual event takes place in early February on a Saturday. Visit www.brandywinewatershed.org for details.

Chapter 22: Legacies of Pride and Shame (March)

Map It! Locate Rainbow Elementary School in Coatesville.

Map It! Locate the historic Lukens district and future National Iron and Steel Heritage Museum on 1st Avenue in Coatesville.

Check It Out! Research the history of the Lukens family and the development of the iron and steel industry in Coatesville. Be sure to check out Rebecca Pennock Lukens. See Appendix M to try to fill out the worksheet before reading further.

Map It! Locate the Saalbach Farm, a Brandywine Valley Association property, on Bonsall School Road between Routes 340 and 30, in West Caln Township, just north of Coatesville.

Check It Out! Look up various birds of the area. Or go on a bird walk and look for different kinds of birds.

Check It Out! Research the calls of frogs and songbirds. Use a Birdsong Identiflyer (for information or a demonstration, visit www.identiflyer.com) that has both bird and frog calls.

Check It Out! Research the life cycle of the dobsonfly, whose larval form is known as a hellgrammite. Find out why its nickname is "toe biter."

Check It Out! Research the lynching of Zacharia Walker, the last lynching in Pennsylvania. Visit http://www.pa-history.org/publications/pahistory.html# and go to Volume 54, Number 2 (April 1987) for the article "A Crooked Death: Coatesville,

Pennsylvania and the Lynching of Zacharia Walker," by Raymond M. Hyser and Dennis B. Downey.

Field Trips! Visit Lukens National Historic District/National Iron and Steel Heritage Museum in Coatesville. Visit the Saalbach Farm for a nature program; go to http://www.brandywinewatershed.org/2008/education/school_programs.asp for more information.

Chapter 23: Earth Day Birthday (April)

Map It! Locate Maysie's Farm, just southwest of the intersection of Routes 100 and 401 in West Vincent Township, just north of Marsh Creek State Park.

Check It Out! Research different kinds of dogs and the jobs they have been bred to do.

Check It Out! Research different kinds of vegetables and taste them!

Check It Out! Research the benefits of Community Supported Agriculture in your area and find local farms that offer organic, local goods. Some, like Maysie's Farm, offer educational programs on sustainable food production. For a listing of CSAs in Chester County, visit http://www.chesco.org/agdev/site/default.asp and go to "view food guide."

Try It! Visit http://www.populationeducation.org/media/upload/earthappleelem_nov2002.pdf for a description of this apple demonstration

Try It! Replicate the demonstration that Sam does with Wyatt and Danni. See which setup releases the least amount of soil into the water. For help with the setup, see Appendix N, Till the Hill.

Check It Out! Research different kinds of frogs and their life cycles, including when egg laying and hatching take place. For the northeast, check spring peepers, green frogs, and bullfrogs.

Map It! Locate Marsh Creek State Park in Upper Uwchlan Township.

Rap It! Try singing Itsey Bitsey Spider, rapper-style. Research spiders and add more verses or create poems or another song for any creatures.

Try It! Measure how far you can jump. Then check your height and measure 20 times your height to compare your jumping skills to those of grasshoppers.

Try It! Adapt-A-Body: Try finding everyday items that simulate frog adaptations. You can use goggles, flippers, a party blower (for the tongue). Have a costume contest or a fashion show. You can do this for frogs or any other creature identified in the novel to gain an appreciation for their unique adaptations. See Appendix O.

Check It Out! Find out more about Brandywine Valley Association's Red Streams Blue program. You may even want to volunteer some time for a stream restoration project. For more information, go to www.brandywinewatershed.org.

Check It Out! Research the history of Milford Mills, the community that was flooded to create Marsh Creek State Park. Two books you can get from the library are *The Story of Milford Mills and the Marsh Creek Valley*, by Catherine Quillman, or *The Upper Uwchlan-A Place Betwixt and Between*, by Estelle Cremers and Pamela Shenk.

Check It Out! Learn to canoe. Canoes can be rented at Marsh Creek State Park (lake) or Northbrook Canoe Company (creek). Be sure to check out signs and treatment for hypothermia.

Field Trip! Visit a CSA for a tour or program. Visit Marsh Creek State Park.

Chapter 24: Breakthrough (May)

Sticker It Red or Blue! Place red or blue stickers on each of the sites described in the chart.

Try It! Try saying these Lenape words. For help with pronunciation, visit http://www.talk-lenape.org/index.php and search for the pronunciation of the English words "good" and "thank you."

Chapter 25: Sweet Waters (May)

Map It! Find Reeceville Elementary, Friendship Elementary, and North Brandywine Middle Schools in West Brandywine Township, northeast of the intersection of Routes 30 and 82.

Check It Out! Listen to the call of the mourning dove; do an online search or use a Birdsong Identiflyer (for information or a demonstration, visit www.identiflyer.com).

Check It Out! Locate the highpoints between the East and West Branches of the Brandywine at this location and/or follow the tributaries near this location to the Brandywine.

Map It! Locate Hibernia County Park.

Check It Out! Research the snapping turtle's life cycle, including habitat and reproduction.

Map It! Locate Chambers Lake and the confluence of Birch Run and West Branch Brandywine, both in Hibernia County Park north of Coatesville.

Check It Out! Research how the iron mill functioned, based on water power, at Hibernia.

Check It Out! Learn about the Amish and Mennonite communities in Chester and Lancaster Counties.

Map It! Locate the West Branch Brandywine at Suplee Road, just east of the borough of Honey Brook.

Rap It! You know what to do! See Appendix F.

Check It Out! Research BMPs or Best Management Practices for farms. Demonstrate BMP practices using an Enviroscape model, which may be available for loan through soil conservation and water protection agencies, including Brandywine Valley Association.

Map It! Locate the East Branch Brandywine at Suplee Road, east of the West Branch Brandywine and Suplee Road site.

Map It! Locate Struble Lake in Honey Brook Township. Note how its waters go to the East Branch of the Brandywine.

Map It! Locate West Nantmeal Township.

Map It! Locate Langoma, now known as St. Mary's of Providence.

Check It Out! Research life at a typical iron mill in the 1800s.

Map It! Locate Springton Manor Farm and Indian Run.

Check It Out! Research the life cycle of butterflies.

Check It Out! Research the life cycle of the spittlebug.

Check It Out! Research and learn to identify poison ivy.

Check It Out! Research and learn to identify jewelweed. In late summer, find out why it's also known as "Touch Me Not."

Try It! Explore the life under a rotting log. Try to identify the critters you find. If you're lucky enough to find a large black, orange, and yellow millipede, smell it.

Check It Out! Research the crane fly. Be sure to see pictures of its larval stage and its adult stage.

Field Trips! Lots of field trip possibilities are in this chapter: Hibernia County Park (Chambers Lake, Hibernia Mansion, Iron Mill Ruins); Honey Brook; Struble Lake; St. Mary's of Providence mansion; Barneston Dam; Springton Manor Farm (Butterfly House).

Chapter 26: Thumbs Up (June)

Try It! Check out International Clothesline Week in early June and try using a clothesline instead of a clothes dryer.

Sticker It Red or Blue! Place red or blue stickers on each of the sites described in the chart.

Map It! Observe the entire map and look for the areas of red stickers and the areas of blue stickers. Try to draw correlations between the sticker colors and the nearby land uses. Streams running through boroughs, near major roads, and through open, unprotected farmland should be clues.

Chapter 27: Tennis, Anyone? (June)

Map It! Locate Honey Brook.

Chapter 28: The View (June)

Check It Out! Research landfills and how they function.

Map It! Locate Lanchester Landfill in Honey Brook Township near Route 322 and the boundary between Lancaster and Chester Counties (see where the term "Lanchester" came from).

Field Trip! Schedule a guided tour of the Lanchester Landfill (also known as Chester County Solid Waste Authority), including the scenic overlook. Get information at www.chestercountyswa.org.

Appendix B

DR. FLO'S CHOCOLATE CHIP OATMEAL COOKIES

Makes about 2 dozen cookies

Ingredients:

- o 2 cups flour
- o 1 cup rolled oats
- o 1 teaspoon cinnamon
- o ½ teaspoon baking soda
- o 1 stick of butter (8 tablespoons)
- o 1 ½ cups of packed dark brown sugar
- o 2 large eggs
- o 1 teaspoon vanilla extract
- o 2 cups (at least!) semisweet chocolate chips
- o 1 cup of chopped pecans
- o ¾ cup of raisins

Directions:

1) Preheat oven to 375 degrees. Position racks in upper and lower thirds of oven.
2) In a large bowl, combine flour, oats, cinnamon and baking soda. Set aside.
3) In a medium saucepan, melt butter over medium heat. Remove from heat and stir in brown sugar, eggs and vanilla.
4) Add flour mixture to butter mixture and stir until well-combined.
5) Stir in chocolate chips, pecans and raisins.
6) Refrigerate dough for one hour.
7) Line 2 baking sheets with parchment paper and drop spoonfuls of dough about 1 inch apart.
8) Bake for 13-15 minutes, rotating pans halfway through cooking, until bottom of cookies are golden brown.
9) Enjoy cookies while reading Sweet Water Hunt!

Appendix C

PROCEDURES AND BIOTIC INDEX CHART FOR STREAM STUDY

Stream Study Procedures: Biological Assessment

Objective: To observe the macroinvertebrates that comprise the stream bottom to determine the quality of the stream water.

Time Needed: Approximately 45 minutes.

Materials Needed: Empty, clean yogurt cups, paint brushes, plastic spoons, shallow white pans, magnifiers, Biotic Index chart, pencils, stream insect guide books, 1-meter-square seine net (optional).

Procedures:

1. Before proceeding to the stream, discuss the concept of a stream community by asking the students what they think they will find in the stream. Have them name organisms until they include some plants and insects.
2. Define the term "Community" and identify the plants and animals they named as producers, consumers or decomposers.
3. Have the students describe a stream as if they were explaining it to someone who had never seen one before. (Freshwater that moves downhill.)
4. Upon approaching the stream, listen for sounds and look for tracks to identify some of the stream inhabitants/users.
5. Stream Study:

   a. Define boundaries of the stream study, distribute equipment and let students explore.

b. If the students do not have a seine net, have them lift rocks and gently brush macroinvertebrates from the rocks into cups of water. Sort the macros; then proceed to c. v. below.

c. If the students do have a seine net, follow these procedures:

    i. Have students place the seine net in the water facing upstream. Line the bottom of the net with rocks to hold down the net so nothing floats under it. Make sure the students do not tip the net so that water flows over its top.

    ii. While several students hold the net in place, have other students pick up and scrub rocks in the square meter upstream of the net so that anything they scrub off of the rocks flows into the net.

    iii. When the entire square meter area had been scrubbed, have the students remove the rocks on the bottom of the net and lift the net carefully so they don't lose anything caught in the net.

    iv. Have the students place the net on dry land. Have them gather around the net and sort macroinvertebrates from the net into cups of water, gently picking them up with paint brushes.

    v. After all the macroinvertebrates have been sorted, classify them using the Stream Study chart in this appendix and determine the water quality.

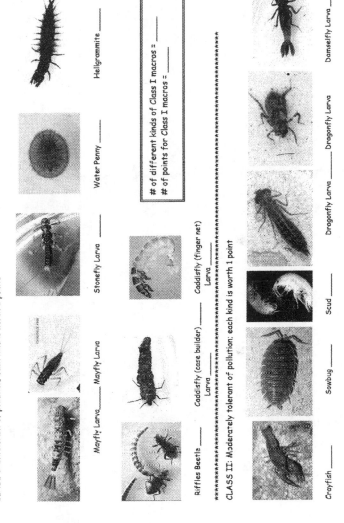

STREAM MACROINVERTEBRATE SIMPLE WATER QUALITY ASSESSMENT

****

CLASS I: Intolerant of pollution; each kind is worth 2 points

Mayfly Larva_____   Mayfly Larva_____   Stonefly Larva_____   Water Penny_____   Hellgrammite_____

Riffles Beetle_____   Caddisfly (case builder)_____   Caddisfly (finger net)_____
                        Larva   Larva

# of different kinds of Class I macros = _____
# of points for Class I macros = _____

****

CLASS II: Moderately tolerant of pollution; each kind is worth 1 point

Crayfish_____   Sowbug_____   Scud_____   Dragonfly Larva_____   Dragonfly Larva_____   Damselfly Larva_____

STREAM MACROINVERTEBRATE SIMPLE WATER QUALITY
ASSESSMENT (SIDE W/CLASS I)

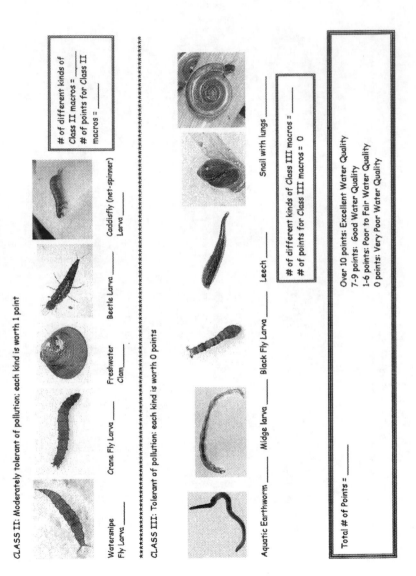

CLASS II: Moderately tolerant of pollution: each kind is worth 1 point

Watersnipe Fly Larva _____   Crane Fly Larva _____   Freshwater Clam _____   Beetle Larva _____   Caddisfly (net-spinner) Larva _____

# of different kinds of Class II macros = _____
# of points for Class II macros = _____

****

CLASS III: Tolerant of pollution: each kind is worth 0 points

Aquatic Earthworm _____   Midge larva _____   Black Fly Larva _____   Leech _____   Snail with lungs _____

# of different kinds of Class III macros = _____
# of points for Class III macros = 0

Total # of Points = _____

Over 10 points: Excellent Water Quality
7-9 points: Good Water Quality
1-6 points: Poor to Fair Water Quality
0 points: Very Poor Water Quality

STREAM MACROINVERTEBRATE SIMPLE WATER QUALITY ASSESSMENT (SIDE W/CLASS II)

Appendix D

STREAM STUDY:
PHYSICAL AND CHEMICAL
TESTING OF THE CREEK

Overview:

In water-quality assessment, the most important test is the biological test, in terms of its macroinvertebrate population/diversity. However, to understand the results of a biological test, it is helpful to understand the factors that influence that biology, including the physical attributes of the stream and the chemical influences. The physical parameters of a stream can influence the chemistry that, in turn, affects the biology.

The physical attributes of the stream include water temperature, turbidity, and velocity; all affect the chemistry of the stream. For example, a slow-moving, smooth area of a stream will have a lower dissolved oxygen content than a fast-moving, rocky area. A colder area of a stream will contain more oxygen than a warmer area. A cloudy area of the stream will absorb more heat than a clear area.

The chemistry of the stream affects the biology. Different biological organisms require different amounts of oxygen and can tolerate different ranges of pH as well as other chemicals in the water.

When studying a stream, it is essential that the biology test is conducted, and very helpful to conduct the physical and chemical tests as well, since all three areas are interrelated.

Physical Testing:

The purpose is to assess the physical attributes of the test site, as they may help explain the chemistry and biology. It is also helpful to assess the physical attributes to ensure comparability from one test site to the next. Physical tests include: temperature, turbidity, and volume of flow (which is determined by width, depth, velocity, and type of stream bottom).

Background:

- Water temperature affects dissolved oxygen content: the colder the water, the more dissolved oxygen it can contain. Organisms in the stream have different oxygen requirements.
- Turbidity measures the clarity of the water: the clearer the water, the colder it can be. Murky water absorbs sunlight and causes an increase in water temperature, decreasing oxygen content.
- Width, depth, velocity, and stream bottom determine volume of flow. The most important test here is the velocity, which affects oxygen content. The faster and more turbulent the water, the more oxygen it can contain.
- Determining volume of flow is great for comparing areas upstream and downstream. Volume of flow is the amount of water that flows through a point in the creek at a given time. At a particular test site, it is measured as: average width multiplied by average depth multiplied by average velocity multiplied by a stream bottom constant. The stream bottom constant is .8 for a rocky bottom; .9 for a smooth bottom.

Average Width X Average Depth X Average Velocity X Stream Constant = Volume of Flow

Testing:

Temperature:

- Materials Needed: Thermometer
- Procedures: Measure both air and water temperature. For any temperature, place the thermometer in the spot and leave undisturbed for a minute. Temperatures may need to be taken in shady and sunny spots and averaged.

Turbidity:

- Materials Needed: Turbidity tube, clean yogurt cup
- Procedures:

  o Using a yogurt cup, scoop water into the turbidity tube until the black and white graphic at the bottom can no longer be distinguished when peering into the tube.
  o Record the centimeter reading at the water line.

Width:

- Materials Needed: Long rope and meter stick OR long measuring tape (long enough to go the width of the stream)
- Procedures:

  o Secure one end of rope to far bank of stream.
  o Stretch rope taut across stream.
  o Secure other end of rope to close bank of stream.
  o Using the meter stick, measure the meters along the rope from bank to bank.
  o Record results of width.
  o Repeat for widest area of stream and narrowest area of stream in the general area tested. Average the widths and record.

Depth:

- Materials Needed: Meter stick
- Procedures:

  o Approximate 4 equidistant points along the width of the stream.
  o At each of the 4 points, measure the depth of the stream. To do so, turn meter stick so that it "slices" the water. Make sure that no one upstream is interfering with water flow.
  o Record depth at each of the 4 sites.

- o Average the 4 depths by adding them all and dividing the sum by 4. Record. (Depth 1 + Depth 2 + Depth 3+ Depth 4 / 4 = Average Depth)

Velocity:

- Materials Needed: meter stick, tennis ball or cork or other floating object, stopwatch
- Procedures:

  - o At the 4 points where depth was measured, conduct the velocity test. At each point, run the velocity test 3 or 4 times and determine an average.
  - o Student #1: Hold meter stick parallel to stream flow and call out start when ball reaches top of meter stick and stop when ball reaches end of meter stick.
  - o Student #2: Time the test with stopwatch.
  - o Student #3: Record the results in seconds per meter.
  - o Student #4: Stand about one meter upstream of meter stick, but not blocking the stream flow; drop the tennis ball into stream so that it flows downstream parallel to length of meter stick without making contact with meter stick (this may take some adjusting).
  - o Student #5: Stand about one meter downstream of meter stick; catch the tennis ball after the velocity test has been run.
  - o For each point, average the velocities.
  - o Average the average velocities and record.(Average Velocity (AV) 1 + AV 2 + AV 3+ AV 4 / 4 = Average Velocity)
  - o Velocities are measured in meters per second. You have measured them in seconds per meter. To address this, simply flip the numerator and denominator. For example, if the velocity was .5 seconds per meter (.5 s/m), flip it to 1 meter per .5 seconds (1m/.5s) and convert (2m/sec).

Volume of Flow:

- Materials Needed: none
- Procedures:

- o To find volume of flow, multiply average width (in meters) by average depth (in meters) by average velocity (in meters per second) by bottom constant (.8 if bottom is rough; .9 if bottom is smooth).
- o AW x AD x AV x bottom constant = Volume of flow (cubic meters per second).

Chemical Testing:

The purpose is to measure basic chemicals in the water to assess water quality and relate to biological results. The most important tests to run are pH and Dissolved Oxygen. Other tests can be run including nitrate and phosphate tests. LaMotte Company and Hach Company sell test kits. It is important to make sure that test kits contain chemicals that have not expired, so periodically check the expiration dates of your chemicals and replenish as needed.

Background:
pH:

- pH is the measure of hydrogen or hydroxyl ions in water.
- pH is a measure of the concentration of acid or base in a solution based on its concentration of hydrogen or hydroxyl ions.
- The stronger the acid or base content, the more corrosive the substance.
- pH is measured on a scale of 0 to 14:

  - o Neutral water has a pH of 7.
  - o Acidic water has pH values less than 7, with 0 being the most acidic.
  - o Basic water has values greater than 7, with 14 being the most basic.
  - o A change of 1 unit on a pH scale represents a 10 fold change in the pH, so that water with pH of 6 is 10 times more acidic than water with a pH of 7, and water with a pH of 5 is 100 times more acidic than water with a pH of 7.

- A pH range from 5.5 to 8.2 is healthiest for diverse aquatic life.

Dissolved Oxygen (DO)

- DO is the measure of oxygen contained in solution of a water sample. It is measured in parts per million (ppm) or milligrams/liter (mg/l). One ppm is like:

  o four drops of ink in a 55-gallon barrel of water, mixed thoroughly
  o one inch in 16 miles
  o one second in 11.5 days
  o one minute in two years

- A good amount of DO for most macroinvertebrates is over 10 ppm or mg/l.
- DO is the most important test that indicates organic pollution.
- DO depends on water temperature: the colder the water, the higher the DO can be.
- DO enters the water by photosynthesis and by the atmosphere, so it changes based on weather, water turbulence, time of day, plant life and animal life.

Testing:
pH:

- Materials Needed: pH test kit
- Procedures: Follow instructions on kit to test pH. Record results.

Dissolved Oxygen:

- Materials Needed: Dissolved oxygen test kit
- Procedures: Follow instructions on kit to test Dissolved Oxygen. Record results.

Stream Study Data Sheet: Physical and Chemical Testing of the Creek

BACKGROUND/BIOLOGICAL INDEX

Date: Time:

Location:

Method of Sampling: seine net sampler   1 meter area of rocks   general area search

(circle one)

Weather Conditions: Weather in Last 24 Hours:

Biotic Index:

Points of Interest (describe the landscape/factors surrounding the study site):

Study completed by:

CHEMICAL DATA

pH = _____   Dissolved Oxygen = _____ ppm or mg/l

****

PHYSICAL DATA

Temperature: Air = _____   Water = _____   Turbidity: _____cm

Average Width: _____ + _____ = _____/2 = _____ meters
                           Wide point   narrow point   total divided by 2

Average Depth (in meters): _____ + _____ + _____ + _____ = _____ /4 = _____ meters
                                         Point 1 Point 2 Point 3 Point 4   total divided by 4

Average Velocity (AV) (in seconds):

_____ + _____ + _____ + _____ = _____/4   = _____ average seconds/meter
 AV 1   AV 2   AV 3   AV 4   total divided by 4 = average velocity in seconds per meter

Convert to meters per second:

Volume of Flow (in cubic meters per second): Average Width _____x Average Depth_____

x Average Velocity_____ x Bottom Constant (.8 or .9)_____ = Vol of Flow _____m3/sec

Appendix E

THE STORY OF OUR TOWN

Objective: To see how land pollutants affect water and add to water pollution

Time Needed: Approximately 15 minutes.

Materials Needed:

- o   Container of water
- o   Stirring spoon
- o   Containers of the following "pollutants": soil; liquid soap; yellow powdered drink mix for fertilizer; salt; trash; pancake syrup for oil; watered-down chocolate syrup for sludge; watered-down chocolate syrup for wastewater; red powdered drink mix for pesticides; chocolate syrup (not watered-down) for manure.

Procedures: Read through "The Story of Our Town" and add pollutants to the container of water to observe how pollutants can add up.

The Story of Our Town:

"Once upon a time, there was a wonderful, pristine town full of industrious, hard-working people. Our Town had lots of houses as well as a few industries in town and was surrounded by lovely farmland where cows and horses grazed the days away. Through the center of town ran a lovely creek that the people called Denial River. The people of Our Town kept their lawns green and neat and their cars washed. Everything looked so beautiful in Our Town that new people moved in to share in the splendor of such an ideal community. The people of Our Town depended on the Denial River for its water supply (point out the water supply) and the people were happy. As time went on and the townspeople worked away, Our Town grew and grew.

New housing developments were built for the growing population, so, of course, some soil started to get washed into Denial River. (Have student with soil pour it into the water source.) The new residents, alongside the others, kept their driveways freshly paved. They washed their cars, they watered their lawns, and they did their laundry diligently. Once in a while, however, a septic system would back up or the soap from a washed car would flow down the driveway into the storm drain and go straight into the Denial River. (Student with liquid soap pours it into water source.)

Many of the people of Our Town enjoyed the sport of golf, but didn't realize how much fertilizer was applied to keep the golf course in town so emerald green and perfectly manicured. (Student adds fertilizer to water source.) In winter, they appreciated how quickly the town's maintenance crew melted the icy roads with salt. (Student adds salt.)

Little did the residents know that there were some troublemakers in Our Town. A few high school seniors didn't think twice when they pitched their papers out the window of their cars on the last day of school, finally feeling liberated from the last twelve years of schoolwork. (Student adds trash to water supply.) The devious Mr. Shady, owner of Shady Auto Repairs, was secretly dumping used motor oil down the storm drain to avoid the expense of legally disposing it. (Student adds oil.) And Ms. Dye, the CEO of Dye Paper Manufacturing, the community's major employer in town, wasn't always exactly in compliance with the state's pollution regulations. So, on occasion, some incompletely-treated industrial waste was dumped into Denial River. (Student adds sludge.) But these characters were few and far between and the people of Our Town didn't really notice what was taking place.

The years passed and the people of Our Town spent them like people everywhere. They watched the rains of hurricane season every fall and had Super Bowl parties every winter, not realizing the overload that these events placed on Our Town's wastewater treatment plant. There was so much overload at times that the plant couldn't process the wastewater completely before returning it to Denial River. (Student adds wastewater.) They drove out to the countryside to view the acres of farmland, not realizing that pesticides had been applied to grow the strawberries and apples they picked. (Student adds pesticide.) And they'd stop in the summer to watch the cows and horses swat flies, but would look away as the animals relieved themselves in the river or on the picturesque farmlands, the products of which would later get washed into Denial River. (Student adds manure.)

Yes, the people of Our Town were happy, until one day they noticed that their river was filthy. Alarmed, they met in the town hall. Mayor Sigmund opened the meeting by pounding his gavel. 'Here, here,' he exclaimed. 'Denial is not just a river in Our Town.'"

Appendix F
SCAT RAP

CHORUS (PG-13 Version)
It starts with an "S"
Ends with a "T"
Comes out of you
And it comes out of me
I know what you're thinking, but don't call it that.
Be scientific and call it scat.

CHORUS (G Version)
It starts with an "S"
Not with a "P"
Comes out of you
And it comes out of me
If poop is what you're thinking, don't call it that.
Be scientific and call it scat.

Verses:
Walking through the woods, your nose goes "ooh"
You know some critter's scat's near you
It may seem gross, well that's O.K.
They don't have toilets to flush it away.
Now don't go scream and lose your lunch
If you look at it closely, you'll learn a bunch.

If you wanna find out what animals eat
Take a good long look at what they excrete
In the scat are all the clues
Parts of their food that their bodies can't use

Park your car in a woods or a field
You might find scat on your windshield
If the berries are purple and white
You just got bombed by a bird in flight

It tells us what they eat
And tells us who they are
And that's what we know about scat so far
So if you wanna find out what critters are around
Take a good long look at the scat on the ground.

Appendix G

DIRTY DEEDS

Objective: To demonstrate some of the techniques used to clean water and how difficult it is to get dirty water clean again.

Materials Needed:

- o   Container of polluted water
- o   Bucket of soapy water for cleaning
- o   Bucket for trash

For each group of students:

- o   A cupful of polluted water from the container
- o   Seltzer or other plastic bottle, cut in half so there is a bottom container and a top "funnel"
- o   Several paper towels
- o   Cotton balls
- o   Sand
- o   Limestone gravel
- o   Charcoal
- o   Coffee filter
- o   Several clear plastic cups

Procedures:

1.   Divide students into teams of 3 or 4. Distribute a set of supplies to each group of students, telling them that they are a team that is to attempt to clean a sample of polluted water.

2.   Let the students, as separate teams, try to figure out a way to clean the polluted water sample, using the supplies distributed (a hint is that all of the supplies are helpful and show them how to fit the funnel into the bottle bottom to make a filter).

3.   After a few minutes of filtering, have the teams present their cleanest sample of water. Be sure to have the students pass around the samples to observe and to smell (no tasting!). Have the entire class vote on which sample is the cleanest and compare it to the original polluted water sample. Also compare it to clean water from the tap. Where does it fit in terms of cleanliness?

4.   (Optional) Test the pH of the samples. Discuss the use of limestone to lower pH.

5.   Clean up and recycle as many of the supplies as possible!

Appendix H

Mrs. Bibbo's Time Line

1. Columbus "discovers" America (1492)
2. Europeans colonize America; trade w/Europe, build mills, etc. (1585 & on)
3. Note: William Penn is deeded southeastern PA (1681)
4. "Americans" are taxed by British: taxation without representation (1760's)
5. Boston Tea Party (1773)
6. 1st Continental Congress meets in Philadelphia (1774)
7. American Revolutionary War begins: "the shot heard round the world"/Battles of Lexington and Concord (1774)
8. 2nd Continental Congress meets in Philadelphia/Official Continental Army is formed (birth of the US Army)/George Washington selected as Commander in Chief (1775)
9. Washington wins battles for Boston (1776). :)
10. Declaration of Independence signed in Philadelphia (1776)
11. New York battles: Americans lose. :(Americans retreat thru NJ to PA (1776)
12. Washington's Crossing of the Delaware Xmas night/Battles of Trenton & Princeton (1776-1777); Americans win. :)
13. Battle of the Brandywine (today!!!) (Actually September 11, 1777); Americans lose. :(
14. British occupy Philadelphia/Washington troops spend winter at Valley Forge (1777-78)
15. Battle of Saratoga; Americans win. :) (1777)
16. Battle of Yorktown (British surrender); end of war (1781). :)
17. 1783 Treaty of Paris officially ends the American Revolution. United States of America is official! :)
18. U.S. Constitution ratified (1787)

Appendix I

THE COMMUNITY WATER CYCLE

Objective: To observe the cycle of water from the creek to our homes and communities and back to the creek

Materials Needed:

o   Copies of Community Water Cycle coloring page
o   Blue, light blue and red crayons

Procedures: Have students follow the course of water from stream to filtration plant to homes to wastewater treatment plant to stream, while coloring the water as it moves through the process.

> KEY: Light Blue = used water, not sanitized and not potable
> Blue   = treated water, sanitized and potable
> Red   = wastewater

1. Water is drawn from the creek to the water treatment plant. Color the stream water light blue.
2. Water is processed at the water treatment plant and then sent to our homes via underground pipes and storage tanks. Color the water blue that leaves the water treatment facility and gets piped to the houses, the storage tank and the individual house.
3. We use water in our homes for hygiene, cooking and washing, laundering and toilet flushing. For the individual house, color the water red that leaves the house and goes to the wastewater treatment plant
4. Water is piped from our homes to a wastewater treatment plant to be cleaned. From tank to tank in the wastewater treatment plant, change the color of the water from red to light blue
5. After water is processed through the wastewater treatment plant, it is returned to the stream. Water entering the stream after being processed at the wastewater treatment plant should be colored light blue

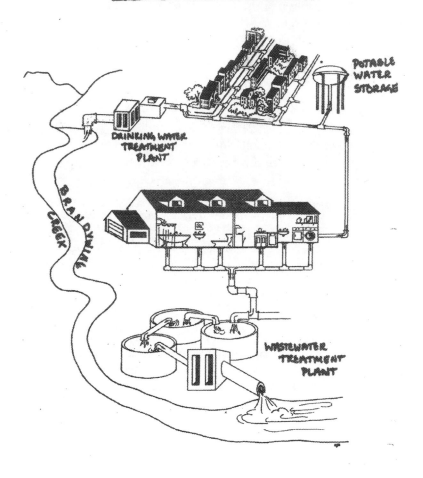

THE COMMUNITY WATER CYCLE

Appendix J

SOIL FILTER

Objective: To demonstrate the effectiveness of the ground for cleaning water

Materials Needed:

o   Soil Filter (see directions): large plastic pretzel container; drill; coffee filter; rocks; stones; sand; dirt; leaves; coffee can; clear, plastic cup
o   Clean cups
o   Water
o   Soil
o   Garden spade

Procedures:

1) Create the soil filter: To create a soil filter, use a large clear plastic pretzel container. Drill a hole in its bottom for water to filter through. Place a coffee filter over the bottom of the container to keep sand from passing through the hole. Layer the soil filter, from bottom to top, with the following items to simulate the ground:

    a) rocks
    b) small stones
    c) sand
    d) dirt
    e) leaves

Place the clear cup in the coffee can and place the soil filter to rest on top of the coffee can.

2) Before demonstrating, prime the filter by pouring clear water into the top so that water runs through the filter. It may take a few minutes for the water to begin to drip through the filter. Make sure the clean cup is in a position to catch the dripping water. After priming, pour out the water from the cup and re-setup.

3) To demonstrate how ground cleans water:

    a. Create dirty water: Mix up soil into water in a cup.
    b. Pour ½ of the dirty water into the soil filter, making sure the filtered water gets caught in the clear cup in the coffee can.
    c. After the water filters through, remove the filter, take out the cup of filtered water in the coffee can and compare with the original dirty water.

Appendix K

RUNOFF RACE

(adapted from WOW! The Wonder of Wetlands, Environmental Concern Inc., P.O. Box P, St. Michaels, Maryland 21663)

Objective: To demonstrate the effectiveness of plantings and wetlands over pavement and grass for cleaning runoff.

Materials Needed:

- 2 equal sized boards, approximately 24" X 18"
- carpeting to cover one board
- 2 catch basins (aluminum or plastic pans)
- container for mixing up dirty water
- spade to dig dirt
- 2 equal sized containers for pouring over boards
- water
- dirt

Procedures:

1) Set up 2 boards of equal size. One board should be bare; the other should have fake grass or carpeting and plant materials covering it.
2) Place the boards on equivalent angles next to each other. The lower end of each board should be placed into a pan or basin to catch the resulting water flow.
3) Mix soil into a sample of water, creating dirty water, to pour over both of the boards. Divide this dirty water into 2 equal containers.
4) At the same time and at the same rate, pour the dirty water onto the top of the angled boards.
5) Observe:
   a) the amount of time it takes for the water to run down each board;
   b) the quality (or dirtiness) of the water in the catch basin;
   c) the quantity of the water in the catch basin.

6) Review how plantings (i.e. wetlands, trees & shrubs) slow down runoff, helping to

   a) reduce flooding in surface waters;
   b) increase infiltration to recharge groundwater;
   c) increase filtration of the runoff, improving water quality of surface waters.

Appendix L

LIMITS OF WATER

Objective: To demonstrate the amount of potable water in the world as a percentage of all of the earth's water.

Materials Needed:

- 1 liter bottle filled to the 1000 ml mark with colored water
- 1 graduated cylinder or other measuring device that can measure up to 29 milliliters of liquid
- 4 clean plastic cups labeled as follows: Polar Ice Caps (22ml); Deep Groundwater (3ml); Polluted Water (1ml); Fresh Water (3ml). Number in parentheses indicates quantities to put into each cup.
- 1 clear unmarked cup
- Globe

Procedures:

1. Display the 1-liter bottle filled with colored water and explain that it represents all the water in the world.
2. Show a globe and ask what 2 kinds of water are on the earth. (Answer: fresh water & salt water).
3. Measure out 29 ml of the water and pour it into an unmarked cup. Ask the students to identify which container holds salt water and which contains fresh water. (Answer: fresh water is the small amount.) Put the salt water away as it is not really useable for us.
4. From the unmarked cup, measure out 22 ml and pour this into the cup marked "Polar Ice Caps." Set it aside indicating that it is not really useable for us since it is locked up in the form of ice.
5. From the remaining 7 ml of water in the unmarked cup, measure out 3ml and place in the cup marked "Deep Groundwater." Explain that this is groundwater that is too deep to be harvested in a cost-effective manner (as opposed to available groundwater for wells). Set it aside.
6. From the remaining 4 ml of water in the unmarked cup, measure out 1 ml and place in the cup marked "Polluted Water." Explain that this fresh water is too polluted to use. Set it aside.
7. Pour the remaining water from the unmarked cup into the graduated cylinder and see how many milliliters remain. (Answer: 3 ml) Pour this into the cup marked "Fresh Water." Explain that, of all the water in the world, this is what we have available for usable fresh water (about .3%).
8. Review the importance of water conservation and pollution control.

Appendix M

LUKENS WORKSHEET

A LITTLE BIT OF LUKENS HISTORY

1. Isaac Pennock owned a mill on Buck Run, a few miles south of the town of _____.

2. In 1794, Pennock and his wife, Martha Webb, had a baby, _____.

3. As a girl, Rebecca learned a lot about the family business of running a mill that produced _____ and _____.

4. When she grew up, Rebecca married a man named _____.

5. Isaac Pennock bought a sawmill in Coatesville from Moses Coates (for whom Coatesville was named). The mill was on the _____ Branch of the Brandywine. Pennock converted the sawmill to an iron mill to make products for the growing railroad and shipping industries.

6. Pennock hired Rebecca's husband, _____, to run the iron mill, Brandywine Iron Works.

7. In 1825, Dr. Charles Lukens died at the age of ____.

8. _____ replaced her husband as the head of Brandywine Iron Works, making her the first female in America to run an industrial company.

9. Her company, later named _____, became a steel mill, specializing in boilerplates used for steam engines for ships and locomotives.

10. Rebecca Webb Pennock Lukens raised ____ daughters to adulthood.

11. Her daughter, Isabella Lukens, married _____

12. When Rebecca Webb Pennock Lukens died in 1854, at the age of ___, her son-in-law took over the Lukens industry.

13. Rebecca is buried in a graveyard in the borough of Ercildoun, just south of _____.

Appendix N

TILL THE HILL

(adapted from "Tilling the Hill" activity from Land and Soil; distributed by the former Pennsylvania Alliance for Environmental Education, now known at Pennsylvania Association of Environmental Educators)

Objective: To demonstrate how different types of land/agricultural practices can affect soil erosion and water quality.

Materials Needed:

- 3 plastic or aluminum trays, about lasagne-pan-sized, with a "V" notch cut out of the center of one end
- wood blocks to elevate one end of each of the pans
- soil to fill each tray
- sod for one tray
- 3 spray bottles filled with water
- ruler to create furrows or terracing
- 3 clear plastic cups to catch runoff

Procedures:

1) Fill each of the trays with moist soil and pack it down. Leave one tray as is; add a layer of sod to one tray; and make contour furrows across the slope of one tray.
2) Set the trays on an equal incline next to each other with the "V" at the lower end of the incline. Place the plastic cups at the "V" to catch the runoff.
3) Simultaneously, spray each of the trays equally with water.
4) After the water washes into the cups, compare the amount of water and sediment in each tray.
5) Review: Which tray lost the most soil? Which tray conserved the most soil? What does each tray represent in terms of farming practices or non-farming?

Appendix O

ADAPT-A-BODY

(inspired by "Adapt-A-Body" activity from What's New at the Zoo, Kangaroo? By Linda Diebert and Andra Tremper, published by Good Apple, Inc. 1982)

Objective: To demonstrate how animals are adapted for survival

Procedures: Using the "accessories" in the table, dress up a student with various adaptations and let the other students guess what organism is represented. Some samples are provided.

Organism	Adaptation	Accessory
Insect	Compound eyes for detection of movement Wings for flying Antennae for non-visual sensing (smell, touch)	Insect "eyes" Dress up, strap-on wings Headband with pipe cleaners
Bat	Membranes between fingers and tail to catch insects and to fly Nocturnal for night hunting Hibernates for conservation of energy in winter Echolocation for detecting prey and avoiding obstacles	Cape Sunglasses Light switch Boomerang
Frog	Webbed feet for swimming Extra membrane over eyes for protection Eyes on top of head for hidden observation Expandable tongue for catching prey Expandable vocal sac (males) for vocalization/attracting mates	Swim fins Goggles Periscope Party Blower Balloon
Opossum	Poor eyesight Poor hearing Pouch for young to develop Prehensile tail Opposable thumbs for grasping Feigns death for defense	Strong glasses Ear muffs Backpack with baby doll Grabber Gloves for feet Tombstone (kick board)
Owl	Talons for catching prey Thickly feathered ankles for protection from biting prey Large eyes for night vision Facial discs for channeling sound to ears, asymmetrically placed Wing feathers with soft fringes for silent flight Poor sense of smell to eat skunks	Garden claws Knee pads or thick socks Binoculars Seltzer bottle cut in half, placed cockeyed around ears Boa Clothespin

Appendix P
Field Guide Example

Common Name: Spicebush
Latin name: Lindera benzoin

Description:
Spicebush, a deciduous shrub growing 6-17 feet, is a member of the Laurel family (Lauraceae) and is easily identified by the spicy scent of its leaves, twigs, and fruit. Dark green leaves are 2-5½ inches long and smooth along the edges, turning bright yellow in the fall. Yellow flowers appear before the leaves in early spring (March-April) in the nodes of the previous year's growth. Female and male flowers occur on different plants (dioecious). Short-stalked fruit mature in the late summer and fall and are shiny red berrylike drupes that enclose a single seed. Spicebush is in the same family as the sassafras tree, but sassafras has leaves that are often lobed.

Habitat:
Spicebush occurs in the understory of rich lowland woods and along streamsides.

Range:
Spicebush is distributed throughout the Eastern United States, from East Texas, Oklahoma, and Kansas eastward to the Atlantic states (although not in Wisconsin) and north to Maine.

Food:
Spicebush leaves and its bright red fruit provide food for more than 20 bird species and other animals (e.g. deer, rabbits, opossums). The spicebush swallowtail (Papilio troilus) lays its eggs on spicebush and other members of the Laurel family.

Life Cycle:
Although the seeds of spicebush are well distributed by birds and other animals, reproduction through root sprouting is common.

Fun Facts:
Spicebush tea can be made from its leaves and twigs and the fruit has been used to make scented sachets. Spicebush extracts have been used medicinally for herbal steams. Because of its rich habitat, early land surveyors used spicebush as an indicator species for prime agricultural land.

Resources Used:
http://plants.usda.gov/plantguide/doc/pg_libe3.doc
http://www.enature.com/flashcard/show_flash_card.asp?recordNumber=TS0383
www.mainenaturalareas.org/docs/rare_plants/links/factsheets/linderabenzoin.pdf

Time Line of Events in the
Brandywine Watershed

1638: Kalmar Nyckel lands at The Rocks (aka Fort Christina) in Wilmington, Delaware, near the mouth of the Brandywine, with the first Swedish settlers. Lenape Indian population was 10-20,000.

1682: William Penn is granted land in the Pennsylvania/Delaware area by the King of England. Penn is the founder of Philadelphia. He was a Quaker, as well as a friend of the Lenape Indians. Sylvania means "forest" or "woods"; thus, Pennsylvania means Penn's woods, which is what the land was comprised of in the 1600s. Lenape Indian population is down to about 2000.

Late 1600s: First mill is built on the Brandywine. Eventually 130 mills are built on the creek.

1703: Downingtown log house is built at the edge of the wilderness.

1706: Indian Rock, on the Brandywine's West Branch, is used to mark the southern boundary of Lenape land that comprises one mile on each side of the creek up to the source.

1720s and 1730s: Lenape Indians complain that mill dams are ruining shad fishing.

1767: Mason-Dixon line is established.

1777: Battle of the Brandywine takes place. British are victorious over the Patriots.

1787: Thomas Gilpin opens Brandywine's first paper mill that later becomes Bancroft Mills, one of the largest textile mills in the world, and now condominiums.

1802: Indian Hannah dies at the approximate age of 71 at the Chester County Poorhouse.

1802: E. I. du Pont builds Eleutherian Mills (gunpowder mills) along the Brandywine (now known as the Hagley Museum).

1810: Isaac Pennock, father of Rebecca Pennock Lukens, establishes Brandywine Iron Works, the predecessor of Lukens Steel Mill and ArcelorMittal.

1825: Rebecca Pennock Lukens takes over Brandywine Iron Works.

1868: Wilmington & Brandywine Railroad is built.

1898: Howard Pyle sets up studio in Chadds Ford gristmill. Howard Pyle is artist and teacher of N.C. Wyeth.

1921: E. I. du Pont gunpowder mill closes.

1945: Brandywine Valley Association is formed.

1961: Chester County Water Resources Authority is established.

1963: Chester County purchases Hibernia land for park.

1964: Brandywine Creek State Park is formed.

1967: Brandywine Conservancy is formed.

1970: The first Earth Day is established, followed by the creation of the Environmental Protection Agency (EPA).

1971: Brandywine River Museum opens, housing many works of the Wyeth family.

1972: The Clean Water Act is established, based on the 1948 Federal Water Pollution Control Act.

1981: Brandywine Valley Association moves to Horatio Myrick's property.

1988: Chester County Parks opens Springton Manor.

1988: Lower Brandywine is designated a scenic river by Pennsylvania Department of Environmental Resources.

1996: Brandywine FLOWS (the Future/the Legacy of Our WaterShed) program is begun at Henderson High School in West Chester.

2006: Brandywine Valley Association implements the Red Streams Blue Program to ensure that all streams in the Brandywine Watershed meet Pennsylvania water-quality standards.

Acknowledgments

This novel has been a labor of love as well as a tribute to the Brandywine Watershed. It is my intention to inspire stewardship of our natural world and all of its inhabitants. This is a work of fiction though it is based on actual watershed data, historical research and folklore.

I am grateful to the following for their help and contributions to this project:

The members of the Sweet Water Hunt Book Club: Abbie, Liv, Elsie, Grace, Hannah, Rema, Max, Meesh, Abby, Sammi, Kelly, Maggie, Journey, Joe and Jackie; and the supporters of the grant: Chester County Conservation District, Downingtown Area School District's STEM Academy and Beaver Creek Elementary School, and Brandywine Valley Association.

Roberta for her encouragement for this project, her never-ending support, her investment of time and energy in this project and her excellent editing skills.

My daughters, Sydney and Eliza, who withstood years of stories and water-related escapades as well as my intrusion into their school lives for the sake of teaching watershed dynamics.

My mother who enthusiastically supported me throughout this project.

The staff at BVA for always supporting my crazy ideas and the EIs, past and present, who always helped with their enthusiasm, generosity of knowledge, and humor.

Kathy, Jane and Bob of BVA's Red Streams Blue program.

Betsy and Ronnie and the girls of GLOW whose support, friendship and kayaking got me through some difficult times.

Zeke and Northbrook Canoe Company for all they do to keep the Brandywine available to all.

Tricia for seeing my potential and helping me move forward.

The Evergreen Foundation, the Jessie Ball du Pont Foundation, and the Pennsylvania Department of Environmental Protection for awarding grants for the programs that led to the creation and piloting of this novel.

Our writer's group of Claire, Jane, Cathie and especially Jan.

Mark & Environmental Ed, whose *Watershed* program remains an inspiration.

Dan, Lynore and all the teachers and students from *Brandywine FLOWS*.

Elizabeth Humphries and Michael Kahn, whose book *Brandywine* provided a wealth of information.

Dorothy Merritts and Robert Walter of F&M, for their work in legacy sediments.

Benny, Max, Phoenix and Jack.

Steve whose success as a psychologist made this project possible.

And, finally, to all the people who work tirelessly to educate our youth and protect the natural resources, cultural integrity and sweet waters of the Brandywine Valley.